U0925010

landscape architecture
风 景 园 林

吉典文化和千朋万友 编
李 壮 主编
时真妹 译

2

大连理工大学出版社

前言

景观设计学（Landscape Architecture）在国外是与建筑学、城市规划相提并论的，而不是从属的学科。

该专业无论在专业起源、学科内容以及位置诸多方面都与国内的风景园林专业很相似。由于文化背景、经济发展速度、社会制度等众多因素不同，国内外园林专业发展水平和侧重都会有所差别。尽管相比之下，国外一些先进国家的景观设计学科专业面更宽、涉足的内容更多，与其他学科渗透的程度更深，但是，专业的性质是相同的，产生和发展的脉胳是一致的。对于其间的差距认识应该是国内园林学科建设和努力学习的目标。

本书汇集了全球众多的景观设计案例，通过大量的照片、文字及分析图将获奖项目原汁原味地展现出来，内容覆盖风景园林实践的各个范围，包括：校园环境、乡村庄园、庭园、传统园林，居住环境、单位园林、景观规划、自然风景、城市开放空间国家森林、国家公园、新城与规划社区、风景车道、娱乐区、自然景观恢复、街景与广场、城市公园、水滨等。按国内设计师的阅读方式整编为六本：1 公园、绿地的景观规划和设计；2 广场的景观设计；3 公共庭院设计、私家庭院设计；4 地球之肺——湿地；5 历史景观保护、改造的规划和设计；6 风景旅游区、度假区的景观规划和设计。案例种类丰富新颖、前沿信息量大，便于读者全方位地参考和解读景观方案的全貌，从这些全球顶级景观新作中获取设计精髓，受益匪浅。

办公广场景观 Office Square Landscape

目 录 CONTENTS

商业广场景观 BUSINESS SQUARE LANDSCAPE

目 录 CONTENTS

休闲广场景观 LEISURE SQUARE LANDSCAPE

办公广场景观
OFFICE SQUARE LANDSCAPE

长庚医院广场
Chang Gung Hospital

LOCATION:
Taipei, Taiwan
AREA:
17.5 ha
PHOTOGRAPHY:
Cheng Chin Ming
TEAM:
mcgregor+partners (landscape architects urban planners)
Ricky Lui and Associates (architects)
DESIGN COMPANY:
McGregor Coxall

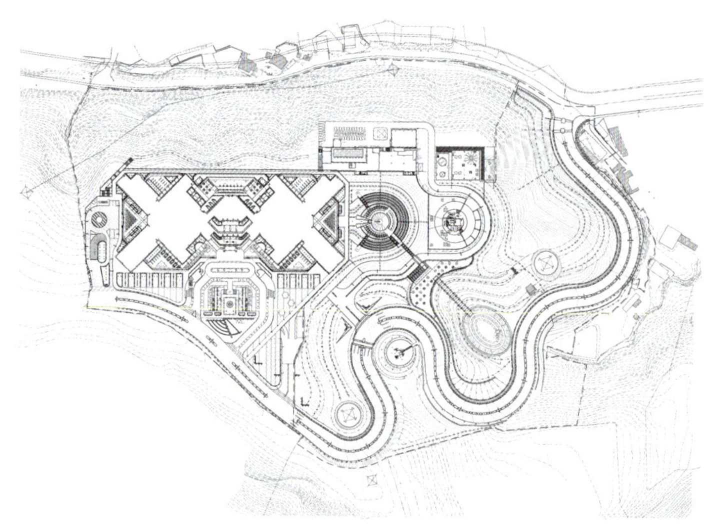

项目地点：
台湾 台北
面积：
17.5 公顷
摄影：
Cheng Chin Ming
团队：
mcgregor+partners (landscape architects urban planners)
Ricky Lui and Associates (architects)
设计公司：
McGregor Coxall

Chang Gung Hospital at the time of the proposed completion was to be the largest hospital in Asia. It was developed on a 17.5ha site by Formosa Plastics, owners of a number of existing hospitals in Taiwan. mcgregor + partners were commissioned to undertake the planning, design, documentation and site review for the projects entire external works. The hospital facility has 3000 beds that service general day surgery and a large percentage of patients who stay for periods longer than three months.

长庚医院，这个由台湾塑胶工业股份有限公司持有，占地 17.5 英亩的医院是完成本设计时亚洲最大的医院。其所有企业，台湾塑胶，在台湾还拥有许多家医院。McGregor 和他的同事们共同完成该项目从计划、设计、搜集资料一直到场地勘察的整个外部工程。医院的 3000 张床位可提供普通的日间手术，也能满足大部分住院 3 个月以上的病人的需求。

The site program was conceived from the thinking that physical and psychological experience of a landscape environment is important in the healing process for patients. Design components include extensive roof gardens, an outdoor cafe, interactive water features, walking trails, sun dial, tea gardens and outdoor art/craft workshop plazas. Storm water from the building is channeled to a thirty–meter long cascading viaduct into an elliptical wetland detention pond for recycling as irrigation.

项目的选址主要来自于对景色环境的生理与心理体验——美景在病人的治疗过程中起着很重要的作用。设计元素包括：开阔的屋顶花园、露天咖啡馆、交互式的水景元素、步行道、日晷仪、茶室、室外艺术 / 工艺作坊广场。建筑物接收的雨水可沿着梯形的高架桥通向椭圆形的湿地蓄水池，可循环做灌溉之用。

美国自然历史博物馆的亚瑟·罗斯台

Arthur Ross Terrace at the American Museum of Natural History

LOCATION:
New York, USA
AREA:
2 acres
DESIGN COMPANY:
Charles Anderson+Partners

项目地点：
美国 纽约
面积：
2 英亩
设计公司：
查尔斯·安德森景观设计事务所

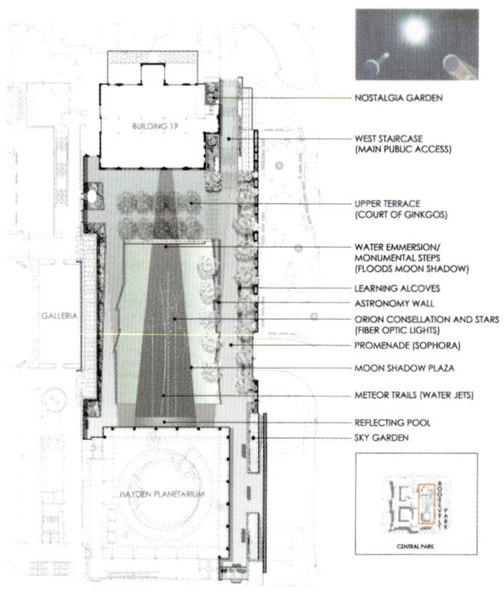

The concept of New York Cityasa make it or break it place often resides in the back of visitors' minds. Its bigness and importance resonates within us as we move about and inhabit spaces in the city. But nowhere is it more felt then when asked to participate in the design and construction of a major new "millennium" addition to one of the city's most important and treasured landmarks - the American Museum of Natural History. Since construction of the Planetarium and new parking garage had already begun, one of the most biggest challenges was to create a design to fit the tree pits and infrastructure of a previous design for the terrace. In retrospect, the process at times felt like attempting to fit a likeness of Julia Robert's face over Walter Mathieu's–a task not easily done.

由于天文台和新停车场的开建，我们所面临的最大挑战之一便是要创建一个能与前次阳台设计中安置的树坑和基础设施相适宜的新设计。如果回顾一下我们这一段阶段的进展，感觉就像是在沃尔特·马修的脸上重塑一张像茱莉亚·罗伯茨的脸。这并不是一项轻松容易的任务。

The rooftop Terrace was designed in tandem with the construction of the Rose and Priest Center for Earth and Space, which replaced the aging Hayden Planetarium, a parking lot and Museum service area.One part of the new construction was the Arthur Ross Terrace, which created an acre of new semi-public open space. The Terrace, a multi-purpose urban plaza constructed over a new parking garage welcomes the public and visitors to the museum,hosts special events, and provides outdoor educational spaces for museum patrons, school children and the general public. The design of the Terrace links the historic, traditional appearance and function of the American Museum of Natural History with the modern design and gleaming materials of the Rose Center for Earth and Space. The physical representation of this concept was inspired by an illustration of the multiple, conical shadows cast by a moon during an eclipse. The futuristic, floating sphere of the Planetarium becomes an eclipsing moon that casts "moon shadows" carrying the sphere's celestial presence across the Terrace.

与屋顶平台设计同步进行的还有玫瑰与牧师中心的建设，此中心用于地球和宇宙展区的展览，代替了原有的老化的海登天文台。除此之外，同时兴建的还有一个停车场和博物馆服务区。新建项目中的一部分便是亚瑟·罗斯台，这个天台为博物馆提供了一英亩见方的半公共露天空地。位于新建停车场上方的天台，将被建成多功能的城市广场，喜迎公众和博物馆参观者们的到来，此外，广场还可用于举办特别活动，为博物馆的赞助商、学龄儿童和广大市民提供户外教育场地。天台的设计融合了美国自然历史博物馆颇具历史意义、传统的外观和功能，以及用于地球和宇宙展区展览的玫瑰中心所采用的现代设计和发光材料。表达这一概念的灵感来源于月食时月亮投射下的多重圆柱形阴影。天文台设计中具有未来主义风格的、浮动的球体就像月食中的月亮一样，在天台上投射下天空中月亮阴晴圆缺变化的阴影。

The Terrace serves as a perfect setting for viewing the exquisite new Planetarium building, a glass cube housing an iconic sphere, which redefined the image of the 128-year old institution while respecting its landmark designation. The Terrace also redefines the landscape vocabulary for the traditional grounds by introducing new materials in a setting of metaphors and simplicity. The Terrace is a space where prehistoric Ginkgo trees meet the stars, creating a place of reflection, learning and respite. Visitors are prompted to contemplate our connection to the earth, as well as explore the wonders of space.

天台还是一处观赏精美绝伦的新天文台的绝佳场所。天文台使用了标志性的星体外罩——晶莹的玻璃立方体。这一设计不仅重新定义了拥有 128 年历史的博物馆的形象，同时还出色地完成了将之变成里程碑的使命。同时，天台的设计还颠覆了传统意义上"风景"这个词汇的概念，引进了一系列具有隐喻性和简洁性的新材料。另外，在天台上，古老的史前银杏树与星际相接，创造了一块可供人们沉思、学习和休憩的场所。在这里，参观者们将会兴致勃勃地研究人类自身与地球的联系，以及探索太空的奥秘。

哥伦比亚广场
The Columbian Plaza

LOCATION:
Columbian USA
AREA:
4 acre
DESIGN COMPANY:
Lango Hansen
AWARDS:
LEED Gold certification

项目地点：
美国 哥伦比亚
面积：
4 英亩
设计公司：
LANGO HANSEN
奖项：
绿色环保认证金奖证书

The Columbian Plaza is situated on a 4-acre site adjacent to a heavily used park. For this highly visited site, Lango Hansen designed a plaza to serve as a forecourt for employees and visitors to the Columbian Newspaper. Special design focus was placed on making the plaza a comfortable, every-day gathering space that is flexible enough to allow for larger, community events throughout the year. Lango Hansen also designed a roof terrace to provide space for larger business or fund-raising events and lunch gatherings. The design used a palette of native and drought-tolerant planting, and an irrigation system that significantly reduced the site’s water demand.

哥伦比亚广场位于一个 4 英亩的地块上，附近是一个使用频率很高的公园。由于这里的使用频率很高，Lango Hansen 为《哥伦比亚报》的员工和来宾建造了一个广场，作为前院使用。设计意在打造一个既可以用于日常聚会、又可以常年用于大型社区活动的舒适场所。Lango Hansen 还设计了一个屋顶平台，为更大的商业或集资活动和午餐聚会提供空间。设计使用了色彩丰富的本地抗旱植物，以及灌溉系统，大大地减少了场地的用水需求。

以色列理工大学的主广场

Main Plaza, Israel Institute of Technology (Technion)

LOCATION:
Haifa Israel
AREA:
4 acre
TEAM:
Shlomo Aronson,
Jorge Salzberg, Ariel Ginat
DESIGN UNITS:
Shlomo Aronson Architects

项目地点：
以色列 海法
面积：
4 英亩
团队：
Shlomo Aronson,
Jorge Salzberg Ariel Ginat
设计单位：
Shlomo Aronson 建筑师事务所

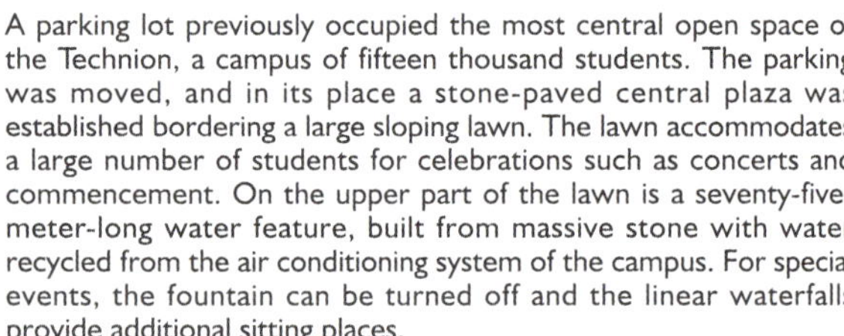

A parking lot previously occupied the most central open space of the Technion, a campus of fifteen thousand students. The parking was moved, and in its place a stone-paved central plaza was established bordering a large sloping lawn. The lawn accommodates a large number of students for celebrations such as concerts and commencement. On the upper part of the lawn is a seventy-five-meter-long water feature, built from massive stone with water recycled from the air conditioning system of the campus. For special events, the fountain can be turned off and the linear waterfalls provide additional sitting places.

以色列理工大学拥有一万五千名学生，它的校园中央有一块空地，空地以前被用作停车场。停车场被迁走后，在这里建成了一个中央广场，广场用石头铺成，并且与一大片草坪坡地相连。在这片草坪上可举办容纳很多学生的大型庆典活动，比如音乐会和毕业典礼。草坪上有一个长达七十五米的水景瀑布设计，这个瀑布通过校园空调系统的水循环从巨石中喷出。在特殊情况下，可以关闭水循环系统，为学生们提供更多的休憩场所。

This area has becomes a favorite place for student's relaxation activities.

这一区域现已成为学生们进行休闲娱乐活动的最佳去处了。

以色列理工大学的主广场

Main Plaza, Israel Institute of Technology (Technion)

LOCATION:
Haifa Israel
AREA:
4 acre
TEAM:
Shlomo Aronson,
Jorge Salzberg, Ariel Ginat
DESIGN UNITS:
Shlomo Aronson Architects

项目地点：
以色列 海法
面积：
4 英亩
团队：
Shlomo Aronson,
Jorge Salzberg Ariel Ginat
设计单位：
Shlomo Aronson 建筑师事务所

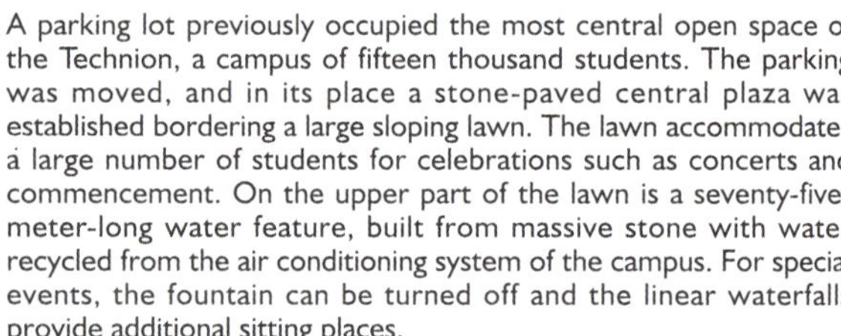

A parking lot previously occupied the most central open space of the Technion, a campus of fifteen thousand students. The parking was moved, and in its place a stone-paved central plaza was established bordering a large sloping lawn. The lawn accommodates a large number of students for celebrations such as concerts and commencement. On the upper part of the lawn is a seventy-five-meter-long water feature, built from massive stone with water recycled from the air conditioning system of the campus. For special events, the fountain can be turned off and the linear waterfalls provide additional sitting places.

以色列理工大学拥有一万五千名学生，它的校园中央有一块空地，空地以前被用作停车场。停车场被迁走后，在这里建成了一个中央广场，广场用石头铺成，并且与一大片草坪坡地相连。在这片草坪上可举办容纳很多学生的大型庆典活动，比如音乐会和毕业典礼。草坪上有一个长达七十五米的水景瀑布设计，这个瀑布通过校园空调系统的水循环从巨石中喷出。在特殊情况下，可以关闭水循环系统，为学生们提供更多的休憩场所。

This area has becomes a favorite place for student's relaxation activities.

这一区域现已成为学生们进行休闲娱乐活动的最佳去处了。

城市沙丘——瑞士 SEB 银行广场

The City Dune —— SEB Bank

LOCATION:
Copenhagen, Denmark
ARCHITECT:
SLA
AREA:
7,300 m^2
DESIGN COMPANY:
SLA

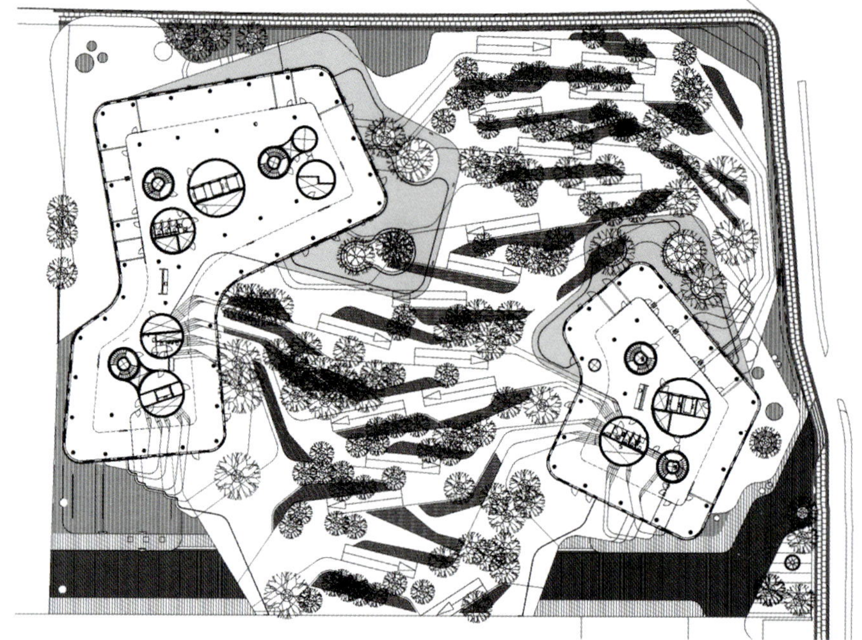

项目地点：
丹麦 哥本哈根
建筑师：
SLA
面积：
7300 平方米
设计公司：
SLA

The City Dune, as the urban space quickly came to be called, is made of white concrete, borrowing its big, folding movement from the sand dunes of Northern Denmark and the snow dunes of the Scandinavian winter. The folding movement and the contour of the terrain not only handle functional and technical demands from drainage, accessibility and lighting to plantation and the creation of a root-friendly bearing layer. It also offers a variety of routes for customers and employees of SEB as well as ordinary Copenhageners, creating an ever changing urban space.

这个城市空间是由白色混凝土浇筑而成的，很快得名为城市沙丘，灵感来源于丹麦北部的沙丘和北欧冬季的雪丘。层层叠叠的混凝土和地形的轮廓线设计不仅是为了排水功能、辅助功能、人工林的可达性和照明功能的需要，它还为 SEB 银行的顾客、员工以及哥本哈根的市民提供了多条线路，以及创建了一片不断变化的城市空间。

To fully experience The City Dune, one has to physically move through it. When passing through the area, the space evolves and opens up in different directions, creating new spatial connections in the process. When ascending from Bernstorffsgade, the space gradually unfolds as you walk along the 300-meter long and winding incline. Looking back against the city, the buildings frame a solid cut of Copenhagen. The ascent from Kalvebod Brygge is shorter and steeper. Here one will soon rise to a splendid view of the harbor.

如果要充分领略城市沙丘，人们一定要亲身去穿越它。当经过该地区时，空间会向不同方向铺开，在展开过程中，创建新的空间联系。当从 Bernstorffsgade 上升时，随着你沿着这 300 米长的蜿蜒斜坡行进，空间会逐渐展开开来。回顾这座城市，建筑物构建了哥本哈根的一个立体的形象标签。Kalvebod Brygge 的上坡路变得更狭窄而且更陡峭了。在这里，人们可以很快上到高处领略到海港的壮丽景色。

The trees and herbaceous borders are placed in fissures between the horizontal planes. Both deciduous and evergreen plantation has been utilized to achieve the metabolism of water throughout the year in addition to enhancing the microclimatic environment with wind and shelter. The trees and plantations are not arranged to emulate nature. It is a new manner of seeing and experiencing nature in the city. The ambition is to create an urban view of nature through a design that clarifies the presence of nature as a process, while simultaneously supporting acclimatization and other functional conditions.

局部下陷的水平面裂缝界定了树木和草坪的边界线。为了达到全年水分的新陈代谢以及改善小气候，在场地上种植了落叶树和常绿树。林地和种植区的设计并不是为了模拟自然，而是在城市里给人一种新的看待和体验自然的方式。这样做的目的在于创造一种城市自然景观，呈现自然过程，与此同时，适宜环境并提供其他的功能。

All in all, The City Dune provides not only acclimatization and utility through the sustainable use of concrete and plantation; it also gives a much needed recreational value to a part of Copenhagen long neglected by city planners.

总而言之，城市沙丘不仅是哥本哈根第一个 100% 可持续地利用混凝土和人工林来营造城市空间的项目，它也提供了被城市规划师长期忽略的非常迫切需要的游憩功能。

戴希曼广场
The Deichmann Square

LOCATION:
Thailand
AREA:
8750 m^2
PHOTOGRAPHER :
Pok Kobkongsanti
DESIGN DIRECTOR :
Pok Kobkongsanti
TEAM :
Kampon Prakobsajakul,Teerayut Pruekpanasan
DESIGN COMPANY :
TROP

项目地点:
泰国
面积:
8750 平方米
摄影师:
Pok Kobkongsanti
设计总监:
Pok Kobkongsanti
团队:
Kampon Prakobsajakul, Teerayut Pruekpanasan
设计公司:
TROP

The Deichmann square and the Negev Gallery constitute a link between Ben-Gurion University campus and the city of Be'er Sheva.

戴希曼广场和内盖夫美术馆将连接本 - 古里安大学校园和贝尔谢巴市。

The square serves as an entrance gate to the western side of the campus, surrounded by existing buildings and the future Negev Gallery. The square offers an outdoor space for cultural and social activities for students and for the city population.

戴希曼广场被周遭现有的建筑环绕其中，日后广场周围还将建立内盖夫美术馆，广场将成为本 - 古里安大学的西门。广场将为学生和城市居民进行户外文化活动、社会活动提供空间。

The square is bordered by the elongated structure of the gallery facing both the city and the campus. Towards the city, the gallery's continuous façade (160 meter in length) unifies the heterogeneous appearance of the existing buildings behind the gallery into a cohesive urban unit. The city façade is accompanied by a sculpture garden creating a green edge to the campus. The two-story high monolithic body of exposed concrete emerges from lawny topography of the northern part of the campus and hovers above an entrance courtyard in the southern part, where it appears to be leaping towards the urban space.

美术馆造型狭长，环绕于广场四周。广场一面面向校园，一面朝向城市。沿城市方向，美术馆绵延的立面（160 米高）将其后高低不平、造型各异的建筑连成一个紧密结合在一起的城市单元。在这道城市立面上雕刻着公园图景，成了大学校园的绿色屏障。裸露在外有两层楼房高的巨大水泥墙体从校园北面的草坪现身，从校园南面的入口操场盘旋而过，仿佛腾空而跃经过下面的城市空间。

The gallery hosts exhibition spaces, museology faculty, workshops and auditorium contributing to the outdoor activities on Deichmann Square. Since the square was designated to accommodate intensive congregation of youth and students, the preferred solution was to allocate limited areas for vegetation. The design of the square with various elements of exposed concrete connects the surrounding buildings both physically and visually, accentuating their common features.

美术馆里设有展览区、博物馆部、工作间和礼堂，与戴希曼广场上的户外活动相辅相成。自从广场开始接纳络绎不绝的年轻人和学生后，人们就倾向于保留下最有限的植被，以创造更多的活动空间。广场在设计上采用了多种裸露的水泥元素，既从建筑材质上也从视觉上将其与周围环境融于一体，增强了共同性。

The square appears as a carpet of integrated strips of concrete paving, vegetation and lighting with concrete benches and trees scattered randomly. The strips of vegetation consist of lawn, Equisetopsida and seasonal plants.

广场看起来像一块由人行路、植被和灯光交织而成的条带分明的地毯，在其上随意浇筑了水泥长椅、栽种着各类树木。条状的植被包括草坪、木贼属和季节性植物。

The first phase to be realized was the Deichmann square to be followed by the Negev Gallery.

首先需要明确的是内盖夫美术馆要在戴希曼广场竣工之后开建。

LG 研究中心静思园广场

LG Research Center Contemplation

LOCATION:
Daeduk, Korea
AREA:
665m^2

项目地点：
韩国 大德
面积：
665 平方米

The granite water wall invites users to engage directly with the water before entering the contemplative area where the pool reflects the movement of the sky. An agitator in the water periodically breaks this mirror. In this spare entrance garden, bamboo, moss, water, and granite are used to create a contemplative courtyard and sculptural pool.

在踏入静思园之前，一道花岗岩水池围墙吸引着使用者去直接感受一下那里的水，园中的水池倒映着天空。水池中的搅动器不时地打碎这面水镜。在这个简朴的入口花园里，有竹子、苔藓、水和花岗岩，共同打造出了一个静思庭院和一个如同雕刻品一般的水池。

LG 研究中心航行园广场
LG Research Center Navigations

LOCATION:
Daeduk, Korea
AREA:
66.5 m^2

项目地点：
韩国 大德
面积：
66.5 平方米

The three granite gateways frame the entry experience of the courtyard. Each holds water troughs which begin the different conditions of water play. The water elements vary, from a reflecting pool to a series of spillways that offer aural and visual interest in this strolling garden.

走进庭院时感受到的就是三条花岗岩通道。每条通道都有水槽，为不同形式的嬉水创造了条件。与水相关的景致形式各异，有倒影池，还有一系列的出水口，这些都使这个散步花园在听觉和视觉上充满了魅力。

LG 研究中心竹园广场
LG Research Center Bamboo Garden

LOCATION:
Daeduk, Korea

项目地点：
韩国 大德

This rooftop courtyard of the LG Corporation in Seoul, Korea consists of geometric arrangements of bamboo that filter light, frame views and provide a lush green backdrop to the activities of the large, open courtyard.

这是韩国首尔 LG 公司的屋顶庭院，是由按几何图形布局的竹子构成的，这些竹子过滤了光线，创造了观景视野，为在这个巨大的开放庭院上的各类活动提供了一个郁郁葱葱的背景。

“河畔折面”——千禧年城市中心广场

Riverside Origami—Millenium City Center

LANDSCAPE ARCHITECT:
Gyorgy Szloszjar, Istvan Steffler, Zoltan Stehli, Edina Csako_Garten Studio Ltd.
ARCHITECT:
Finta Studio Ltd.
AREA:
7,000m²
CLIENT:
Duna Kongresszus Real Estate Development Ltd.
PHOTOGRAPHER:
Garten Studio Ltd., Istvan Steffler

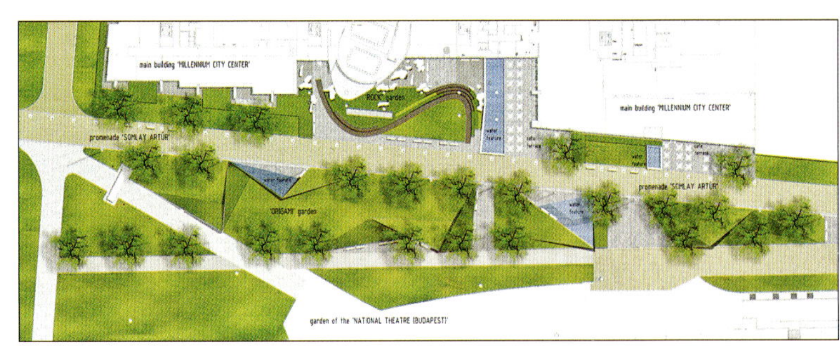

景观设计师：
Gyorgy Szloszjar, Istvan Steffler, Zoltan Stehli, Edina Csako_Garten Studio Ltd.
建筑师：
Finta Studio Ltd.
地点：
布达佩斯，匈牙利
面积：
7，000m²
摄影师：
Garten Studio Ltd., Istvan Steffler

Concept
The main concept of the whole new block was forming a fluent ground floor with public functions, involving the surrounding open space into it. Another intention on higher levels was to divide the longish shape, making structural and visual segmentation and let through the pedestrian cross traffic towards the river and the promenade. On the east and north side of the building there is mostly pedestrian and vehicular traffic space, while on the south side there is the neighboring building site. Public functions like recreation, cafe terraces, sauntering along, having a little piece on a bench or lying on the grass, could be placed on the west side of the building towards river Danube. Garten Studio covered these functions with a frame structure, which can form various spaces while being constant and fluent visually. Another aspect was to design a landscape which has its own ground floor look and gives an interesting sight from above the relatively high office building.

概念：
整个场地的设计概念是形成一个带有公共功能同时也能吸纳周围开放公共空间的流畅性场地空间，另一深层次的意义是划分略长的场地，创造结构和视线上的分区并且方便行人从河对面穿过，在场地中漫步。
场地东部及北部有着大量的步行及车行交通空间，南部毗邻建筑，回廊咖啡厅、坐凳等一些带有公众服务功能的娱乐设施宜安排在建筑西侧靠近多瑙河地带，我们用这种框架结构既满足了空间的连续变化又覆盖了以上的功能需求。另一方面是设计一种既有明显场地形式特点又能从高层建筑上看起来很有趣的景观。

Design

Garten Studio mixed a contemporary design of triangle shape surface forming, with a little origami, which both gave us enough freedom to make the functional spaces and get the unusual, constructed look from above. The roof gardens have a same approach in design, strict grid-like textures, following the raster directions of the house instead of sticking to its outline. The complete project is expected to earn LEED Gold rating, which had effects on design process too. Mostly the adopted technology — rain-age, irrigation, lighting-and material usage-paving materials, plantation had to be considered by LEED requirements.

设计：

我们用现代的设计手法以三角形为基本元素生成整个场地的形式，这又有点像折纸工艺，这样既能满足功能的划分，又给整个设计以不寻常的形式感。屋顶花园的设计也采用类似手法，严格遵从房子栅格的纹理方向而不是其外表皮。节能理念也贯穿着整个项目，项目有望获得能源与环境设计先锋奖的最高级认证。排水、灌溉、照明等先进的技术以及材料和种植技术都是该奖项的要求。

Playa Vista 校园广场

The Campus at Playa Vista

LOCATION:
CA,USA
AREA:
64 ACRE
PHOTOGRAPHY:
Hester + Hardaway, The Office of James Burnett, Dillon Diers, Kerun Ip
DESIGN COMPANY:
The Office of James Burnett

项目地点：
美国 加利福尼亚
面积：
64 英亩
摄影：
Hester + Hardaway, The Office of James Burnett, Dillon Diers, Kerun Ip
设计公司：
The Office of James Burnett

Located on the former site of the Howard Hughes Aircraft facility, the Campus at Playa Vista is a 64 acre office campus that engages the historic industrial fabric of the property. Preservation of the main aircraft hangar, which once housed the famed Spruce Goose, as well as the other ancillary buildings established the LEED accredited architectural and landscape architectural vernacular for the development.

Playa Vista 校园坐落于霍华德休斯飞机工厂的旧址，是一块占地约 26 万平方米的学校办公区，利用了场址上具有历史意义的工业建筑格局。曾经存放著名的 Spruce Goose 飞机的主库房还有其他的辅助建筑被保存了下来，使该地的开发在建筑和景观上具有地域特色，并获得了绿色建筑协会的认证。

A central nine-acre park consisting of sport courts, playground, soccer field, botanical gardens, water features, and a bandshell, serves as the social hub for the campus. Each parcel has park-front access or direct views to the central green providing a strong relationship between architecture and landscape architecture. Richly landscaped courts and roof gardens are integrated in with the proposed and historic buildings providing tenants with easy access to the famed outdoor environment of Southern California.

一个约 36 万平方米的中心公园由球场、操场、足球场、几处花园、几个水景，和一个户外音乐台组成，充当校园的社交中心。每部分空间在公园前侧都有入口，都能直接看到中心的草坪，草坪在建筑和景观之间建立起紧密的联系。景致丰富的庭院和屋顶花园与拟建建筑和历史性的建筑融为一体，为房客提供便利的途径去体会加利福尼亚南部著名的户外环境。

Qu' yvit 雕塑广场
Qu' yvit Plaza

LOCATION:
Antwerp, Belgium
AREA:
200 m²
LANDSCAPE DESIGN:
OMGEVING cvba
CLIENT:
Robelco, city of Antwerp

项目地点：
比利时 安特卫普
面积：
200 平方米
景观设计：
OMGEVING CVBA
委托人：
Robelco，安特卫普市

The artwork Qu' yvit, conceived of by OMGEVING and further developed and designed with the artists' collective Le MuMi, depicts five monumental baobabs to which mythical powers are attributed in African culture. In Africa, they are often a gathering place for activities and a meeting point in the shadow or center of the village. Also on the new Kievitplein, the sculpture attracts business people, residents, children, travelers and tourists and livens up the square so that it forms the focal point of the Kievit neighborhood. The organic structure which is covered with Wisteria's counterbalances the rigid, cold building blocks.

Qu' yvit 雕塑作品是由 OMGEVING 构想的，Le MuMi 艺术团队又做了进一步的创新后最终设计出来，它展现的是五棵巨大壮观的猴面包树，在非洲的文化中，它们被赋予神奇的力量。在非洲，它们位于村子的中心，阴影处常被用来进行各种活动及集会。而且，在新建的 Kievit 广场，这一雕塑吸引了商人、居民、孩童和游客，让这个广场充满活力，成为 Kievit 区域的焦点。这个有机结构覆盖着紫藤，让周围的建筑物显得不那么冰冷死板。

保利国际广场
Poly International Plaza

LOCATION:
Guangzhou China
AREA:
57 HA
TEAM:
I-Hsien Lee, Ye Luo, Aleksandra Dudukovic
DESIGN COMPANY:
SWA Group

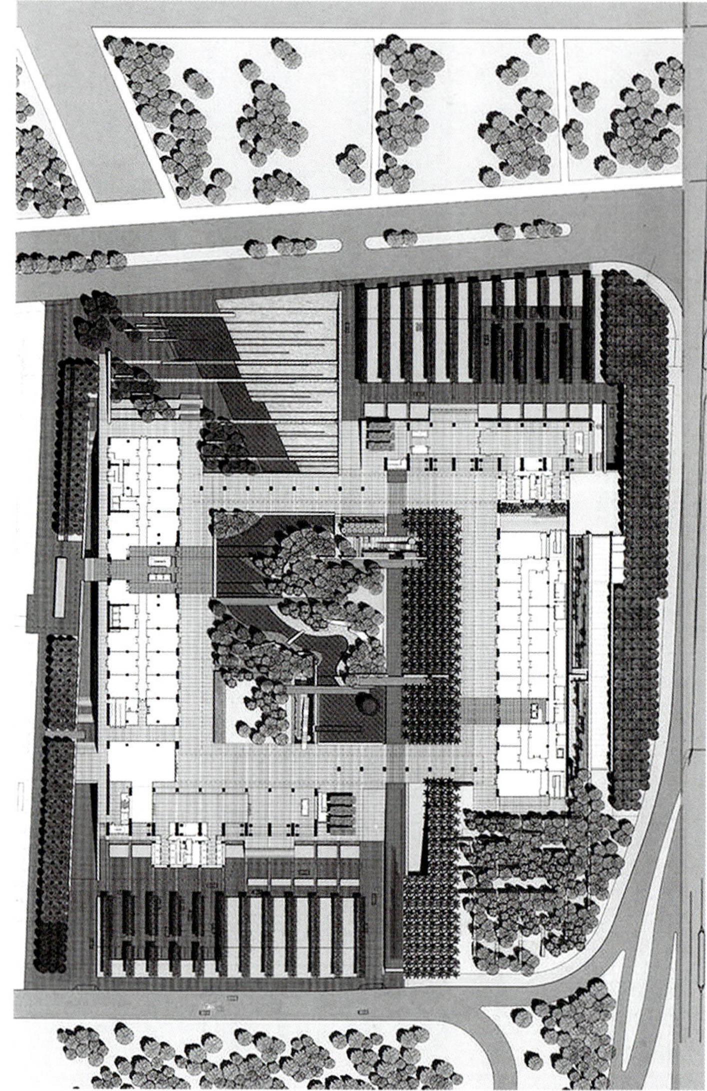

项目地点：
中国 广州
面积：
57 公顷
团队：
I-Hsien Lee, Ye Luo, Aleksandra Dudukovic
设计公司：
SWA Group

Located in Guangzhou, China, the 57 hectare site, which offers 180,000 square meters of office and exhibition space, is part of the city's new exhibition and trade district. Situated between the Pearl River and the historic Pazhou Temple Park, the property was formerly agricultural fields with water channels linked directly to the Pearl River. The context of the river, along with the site's agricultural history, the region's tropical climate, the client's aesthetic appreciation of Chinese gardens, and our overarching aspiration to be innovative, sustainable, and modern have all converged to influence the landscape design.

项目位于中国广州，占地 57 公顷，提供 18 万平方米的办公和展览空间，是该市新的展览及贸易区的一部分。项目场址位于珠江和历史悠久的琶洲庙公园之间，以前是农田，田里有水渠直接与珠江相连。该地段的河景、农业历史、热带气候、客户对中式花园的美学欣赏，还有我们对新颖、持续和现代感的强烈渴望，这些汇集在一起影响了景观的设计。

The architecture and site are designed integrally to embrace habitable green- and sustainable-development strategies. Consisting of two slender north/south facing wafer towers coupled with low-rise podium buildings, the architecture is diagonally offset around a large central garden court. This building configuration and massing capture the prevailing breezes, channeling them through the central garden. As part of a broad architectural plinth, the entire garden court is elevated 1.5 meters to enhance the effects of these breezes. Continuous architectural canopies form an open-air sheltered concourse around the garden's perimeter. To instill a cooling effect, water surfaces are strategically configured to engage these breezes. During the monsoon season, the watercourse serves to partially store and convey storm runoff across the site.

建筑与场地结合在一起设计，体现出宜居、绿色环保和可持续发展的理念。建筑由分别朝南朝北的两栋狭长的高楼组成，伴有底层的裙楼，成对角线围绕在一个巨大的中央庭院周围。这样的建筑布局和体量能招来这儿常吹的微风，并引导风吹过中央花园。整个花园场地属于宽阔的建筑基座的一部分，被抬升了 1.5 米以增强微风吹拂的效果。沿着花园的周边修建了连绵的遮盖，构成了一个户外的遮蔽空间。为了让人体会凉爽的效果，水体区域的布局是经过精心设计的，为了让微风在这里吹拂而过。在雨季，水道还起到储存部分雨水使其在园区内流淌的功能。

休泊克伦广场
Superkilen

LANDSCAPE ARCHITECT:
Topotek 1, BIG Architects, Superflex
COLLABORATION:
Lemming&Eriksson, Help PR&Communication, Heine Petersen
LOCATION:
Nmrrebro, Copenhagen, Denmark
TOTAL AREA:
39,000m^2 (Black Square: 3,700m^2 Green Park: 14,000m^2 Red Square: 9,500m^2)
FUNCTION:
Public Space
AWARD:
1 st Prize
CLIENT:
Copenhagen Municipality, Realdania
PHOTOGRAPHER:
Dragor Luftfoto, Hasse Ferrold, Hanns Joosten, Iwan Baan, Jens Lindhe, Mike Magnussen, Torben Eskerod

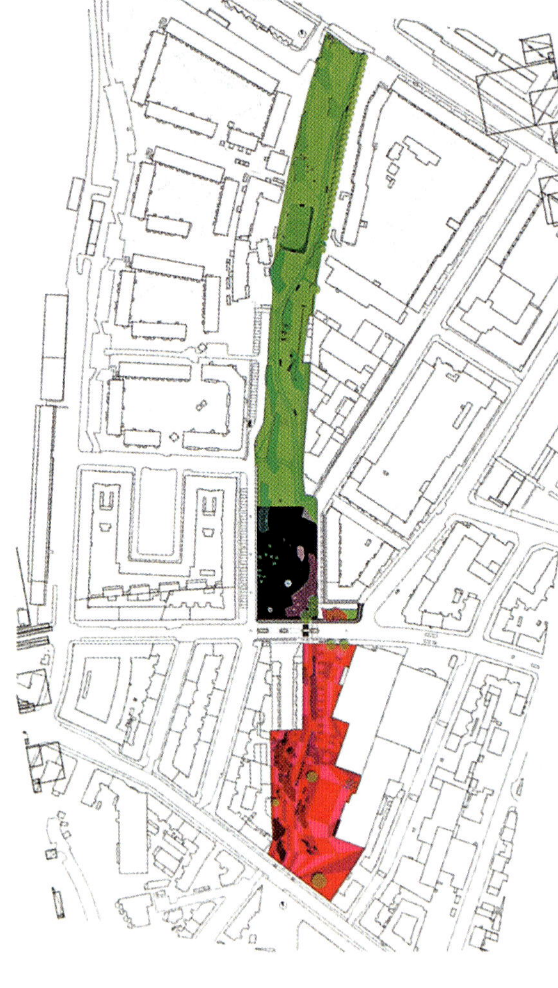

景观设计师：
Topotek 1, BIG Architects, Superflex
合作团队：
Lemming&Eriksson, Help PR&Communication, Heine Petersen
地点：
丹麦哥本哈根，纳赖伯鲁
总面积：
39 000m^2(黑色广场：3700 m^2；绿色广场：14 000 m^2；红色广场：9500 m^2)
功能：
公共空间
所获奖项：
第一名
委托人：
哥本哈根市政府、瑞尔达尼亚基金会
摄影师：
Dragor Luftfoto, Hasse Ferrold, Hanns Joosten, Iwan Baan, Jens Lindhe, Mike Magnussen, Torben Eskerod

Urban Revitalization Superkilen

Superkilen is a heterogenous site-collage in a dense, centrally located neighbourhood in Copenhagen. The strongly international quarter with a mix of different cultures is to be revitalized using open space as a physical framework. This space is to be propelled beyond its current role as a monofunctional transit area into being innovative and dense with synchronicities. Accordingly, the concept aims at enhancing the diverse characters of its protagonists and within the site.

城市振兴之休泊克伦广场

休泊克伦广场位于哥本哈根中心附近，它将很多地点密集拼接在一起。在这个国际区域里不同的文化相互交融，并且利用这个开放空间产生了新的活力。该空间超出其现有的角色，作为单功能中转区域逐渐转变成为创新密集区。相应地，设计的理念是增强这个景点的多样化特征。

© Iwan Baan

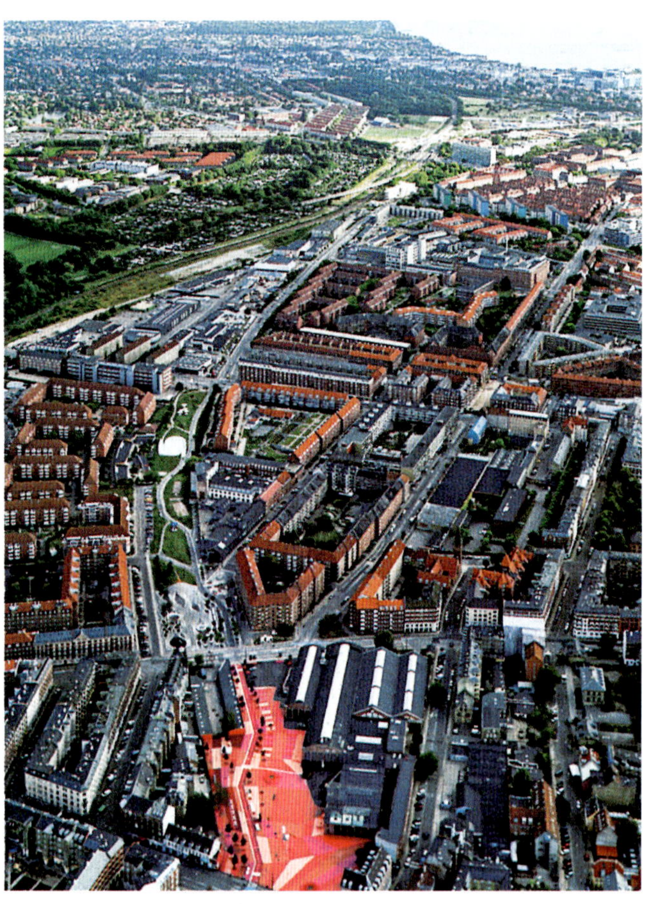

A black square, a red square and a green park will be the matrix of dialogue with the realities of Superkilen. As part of this dialogue, the design reattributes an essential motif from garden-history. In the garden, the translocation of an ideal, the reproduction of another place, of a far-off landscape, is a common theme through time. Where the historic Chinese garden features miniature rock formations of famous mountain ranges, the Japanese zen garden abstracts the sea into waves of gravel. The historic gardens in Florence or Versaille are loaden with allegorical depictions and the historic English landscape garden showcases replications of Greek ruins. In Superkilen this there finds a contemporary, an urban form: a global, universal garden. Here, the transfer of significative elements from other places and cultures reflects the multi-ethnic structure of the neighborhood and activates it.

一座黑色广场、红色广场和绿色广场将成为与休泊克伦广场实际情况进行对话的矩阵。作为对话的一部分，设计从花园历史中提取了一个中心主题。在花园中，理想区域的改变、另一处遥远景观的重现是通过时间考验的常用主题。其中具有历史意义的中国园林突出了著名山脉中岩层的缩影，日本的禅花园以大海为原型，打造了一波又一波的砂砾。通过寓言的方式引进了佛罗伦萨或凡尔赛的历史园林，而英式风景园重现了希腊遗迹。在休泊克伦广场，这一主题展现了当代的城市形态：一个全球性的园林。从其他地方转入的富含意味的元素以及文化都反映了附近区域的多民族结构。

© Iwan Baan

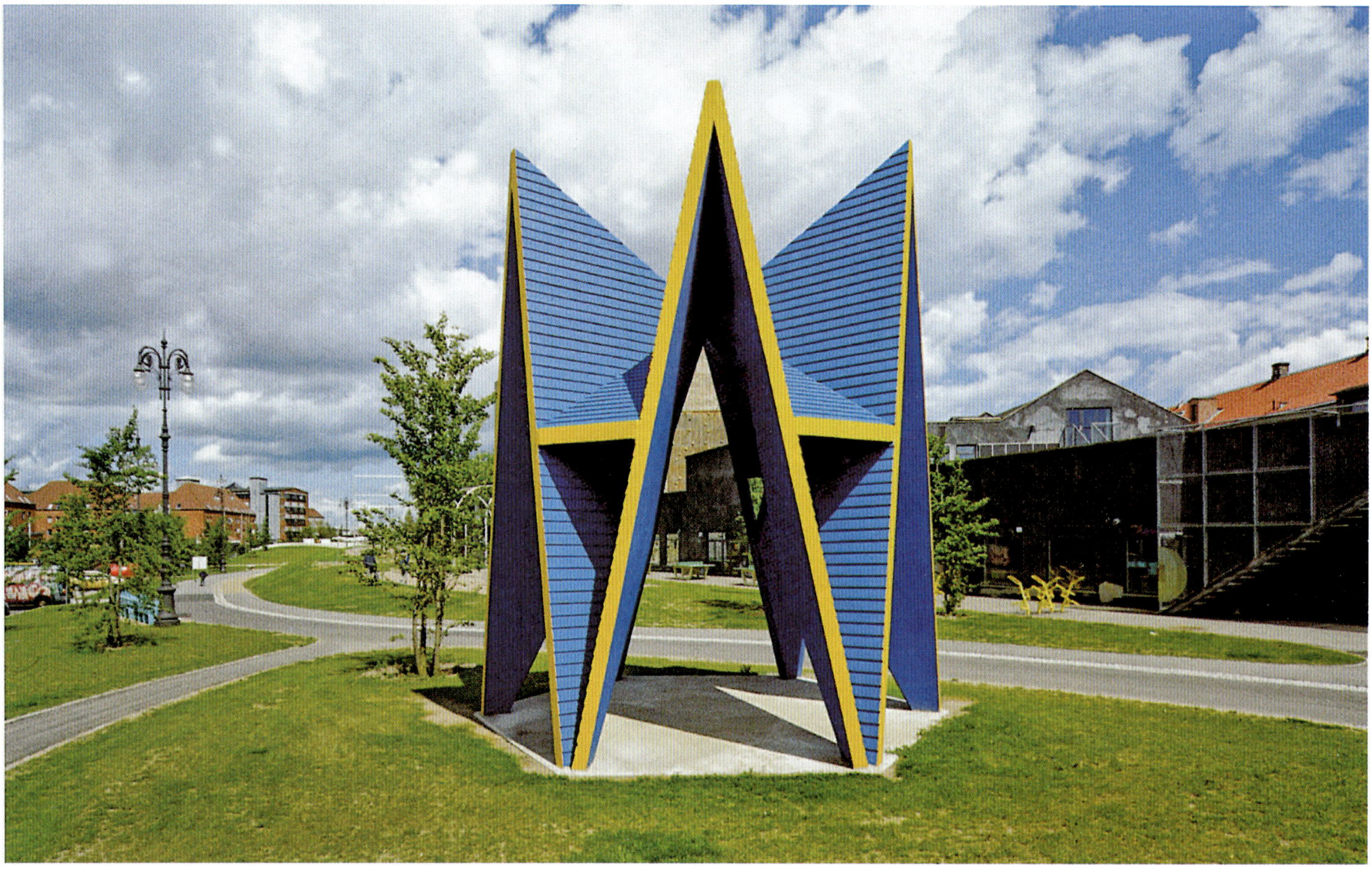

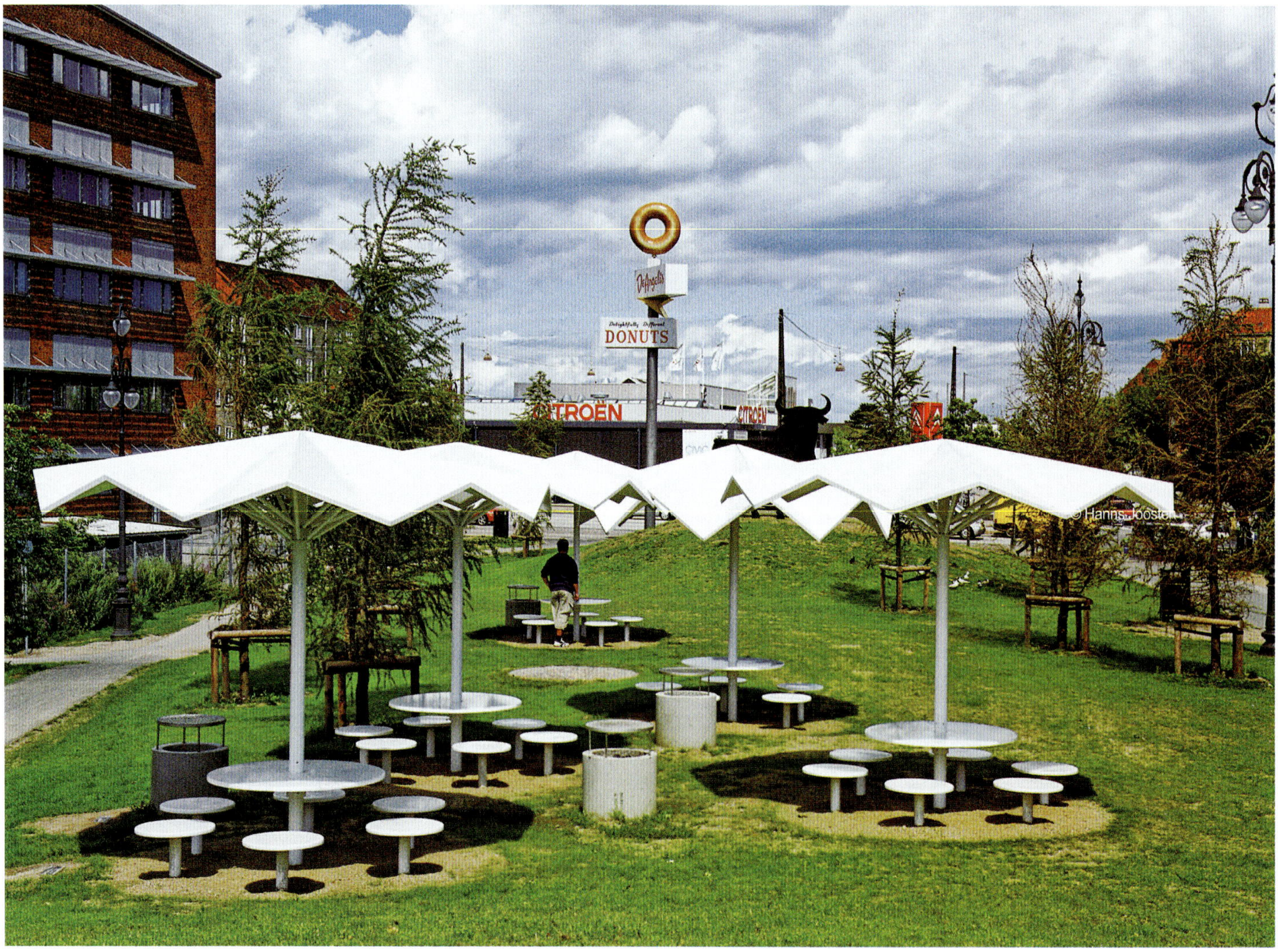
DONUTS
CITROËN
© Hanns Joosten

© Iwan Baan

© Iwan Baan

The furnishing of Superkilen is developed from an international catalogue of urban design elements. In many months of workshops and conversations with residents and local associations the creativity and fantasy of the quarter has been mobilized. Civic participation has been developed as a motor for the design principle of multitude. Round benches, fountains, lamps, fitness equipment and sundry more now projects Superkilen's diversity and international personality onto the matrix of a versatile neighborhood park.

休泊克伦广场的设施源于世界各地的城市设计元素。数月的讲习、与居民间的讨论以及当地集会，共同推动了广场的创造力和吸引力。公民参与已经成为推动多样性设计原理的源泉。圆形的长凳、喷泉、灯具、健身器材以及各种各样的其他物品造就了休泊克伦广场的多样性以及国际特色，它是一个多功能社区公园的原型。

达拉谟学院广场
Durham College Plaza

LOCATION:
ONTARIO, CANADA
IMAGE:
Pete North, North Design Office

项目地点：
加拿大 安大略
图片：
Pete North, North Design Office

With an interior renovation update, the college decided that their concrete slab courtyard could also use improvement. Working with the existing slab and catch basins, three sharply geometric planters were added to fit into the small space. The planters are formed by gabion baskets filled with carefully placed stones to create a sustainable aesthetic, in matching with the college's mandate and program. The height of the gabions allows for sitting, leaning, or placing a laptop to work outdoors. The deep volumes of soil created by the gabion frames are now able to host groupings of multi-stemmed birch trees and periwinkle groundcover, to create a density of green in the small courtyard space. A long bench hovers at one edge, inviting students to enjoy the space. Surrounded by glass, the courtyard, as a summer green oasis or winter spectacle, can also be enjoyed from the updated interior.

随着室内的改造升级，学院决定对混凝土石板庭院也进行改进。保留现有的石板和集水池，把三个几何形状鲜明的花坛融入这个小空间。花坛由里面精心放置了石头的石筐构成，营造出一种持久的美感，符合学院的要求和规划。石筐的高度适合坐着、斜靠或者放一台笔记本电脑进行室外工作。石筐围起的深土种植着枝干繁多的桦树，覆盖着长春花属地被植物，在一个小小的庭院空间创造出浓密的绿色。一条长凳设于一边，邀请学生们来观赏。庭院由玻璃环绕，所以从室内就能欣赏到此地夏日的绿洲或冬天的壮美。

德国广播电台总部
Deutschlandfunk Headquarter

DESIGN COMPANY:
SIC Architectural Design Co., Ltd., Germany
LOCATION:
Germany

设计单位:
德国 SIC 建筑设计责任有限公司
项目地点:
德国

This project includes a 24-story building and a 4-story administration building (with 3-story underground garage). This is a renovation project, so the normal running and broadcasting of the radio station must be guaranteed during the overall design and construction. In addition to the overall renovation and arrangement for the original building, a canteen, an internal courtyard and a main broadcasting center will be built.

本建筑包含一栋 24 层高的楼房以及一栋 4 层高的行政大楼（地下有三层车库）。因为这是一个翻新改建工程，所以在进行整体设计和施工过程的同时，还必须保证电台正常运营以及播放节目。除了对原有建筑进行整体翻新和布局之外，还新建了一个员工餐厅、一个内部庭院以及一个主播放中心。

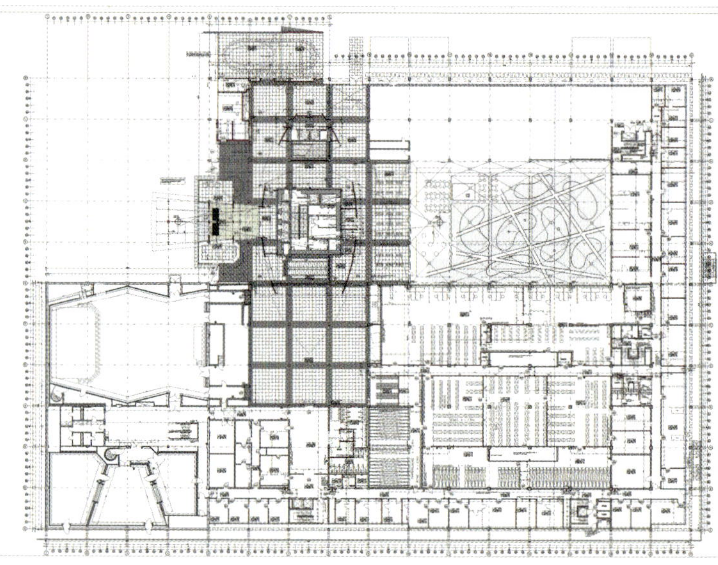

东湖畔花园广场

The Park at Lakeshore East

LOCATION:
ONTARIO, CANADA
IMAGE:
Pete North, North Design Office

项目地点：
加拿大 安大略
图片：
Pete North, North Design Office

The Park at Lakeshore East is a 6-acre urban park that is the central amenity of the 28-acre Lakeshore East development in Chicago ' s Inner Loop. Overlooking the confluence of the Chicago River and Lake Michigan, Lakeshore East is a $4 billion redevelopment that at completion will include 4,950 residential units, 1,500 hotel rooms, 2.2M gross square feet of commercial space, 770,000 SF of retail space and an elementary school.

东湖岸公园是占地约 2.4 万平方米的一座都市花园，是芝加哥内环线约 11.3 万平方米东湖岸开发区域的核心设施。东湖岸区俯瞰芝加哥河和密歇根湖汇合之处，是投入 40 亿美元的重建区，完工后将包括 4950 户住宅、1500 间宾馆客房、20.4 万平方米总值的商业空间、7.2 万平方米的零售空间和一所小学。

Two sweeping promenades serve as the primary circulation across the site and each features a series of fountain basins, seating areas and ornamental gardens. An extension of Field Street ' s axis, the Grand Stair offers a commanding view of the park and accommodates the 25 ' grade differential created by Chicago ' s 3-tiered transit system. Additional amenities include a children ' s garden, dog park and event lawn.

两个宽阔的步行区是贯穿公园最主要的通路，每个步行区的特色是有一系列的喷泉水池、座位区和精美小花园。顺菲尔德大街轴线的一段延伸，叫作华丽阶梯，在那儿可以居高临下俯瞰整个公园，按芝加哥三级运输系统的要求有 25 度的斜坡。其他设施包括儿童花园、遛狗园和举办活动的草坪区。

RENT NOW

甘尼特今日美国总部

Gannett/USA Today

LANDSCAPE ARCHITECT:
Michael Vergason Landscape Architects, Ltd.
L□CATI□N:
Alexandria, Virginia

景观设计师：
Michael Vergason 景观设计有限公司
地点：
亚历山大，弗吉尼亚州

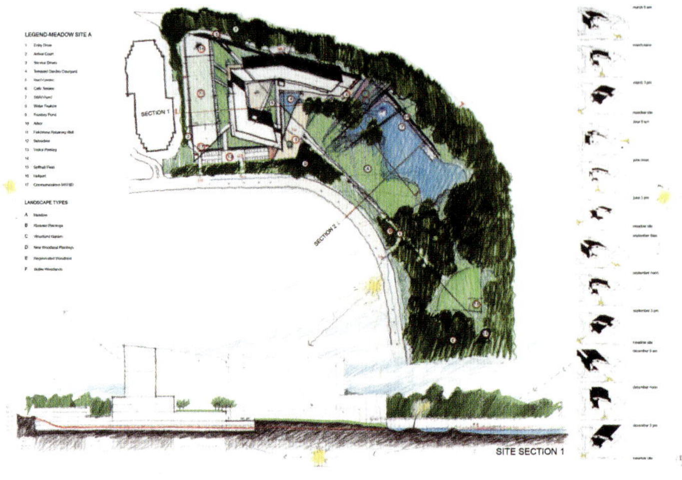

The Gannett/USA Today Headquarter is an ecologically diverse refuge in the rapidly developing business and retail center of Tysons Corner, Virginia. The landscape architect developed a site strategy that seamlessly weaves the indoor and outdoor spaces into a campus of extensive roof gardens and terraces, riparian plantings and preserved woodlands. This project demonstrates the exceptional value of thoughtful site design and site repair for the creation of a distinctive corporate campus.

甘尼特今日美国总部坐落于弗吉尼亚州泰森斯角快速发展的商业和零售中心，是一个生态多样性的保护地。本项目的景观设计师制定了一套方案，把室内和室外空间天衣无缝地编织成为一个带有开阔的屋顶花园、平台、河岸植被和保护林地的园区。这个项目体现了设计师非凡而缜密的场地设计与修复的功力，创建了一个独特的企业园区。

跳跳糖广场
Pop Rocks

LANDSCAPE ARCHITECT:
AFJD Studio, Matthew Soules Architecture
LEAD DESIGN:
Matthew Soules, Joe Dahmen, Amber Frid-Jimenez
WORK TEAM:
Jen Boyle, Byron Chiang, Baktash Ilbeiggi,
Warren Sche ske, DerreckTravis
LOCATION:
Downtown Vancouver, British Columbia, Canada
AREA:
500m^2
CLIENT:
City of Vancouver
PHOTOGRAPER:
Krista Jahnke

景观设计师：
AFJD Studio, Matthew Soules Architecture
主要设计师：
Matthew Soules, Joe Dahmen, Amber Frid-Jimenez
工作团队：
Jen Boyle, Byron Chiang, Baktash Ilbeiggi,
Warren Sche ske, DerreckTravis
项目地点：
加拿大不列颠哥伦比亚温哥华市中心
面积：
500m^2
委托人：
温哥华市
摄影师：
Krista Jahnke

In the spring of 2012 we won a civic commission to design a temporary installation on a prominent city block in the center of downtown Vancouver that sees over 60,000 pedestrians daily. Our mandate was to invent a space for residents and visitors to sit and recline, sunbathe and eat, and interact and play.

在2012年的春天，我们接受委托，在每天有6万行人穿梭其间的温哥华市中心主要街区暂设一个工程。我们的委托内容是为居民和游客提供一个坐躺、沐浴阳光、吃饭、交流和玩乐的地方。

While the strategic use of resources permeates all our design activities, the temporary nature of this commission highlighted the necessity for a profoundly efficient use of resources. The project is fabricated entirely from post-consumer and post-industrial waste from the metropolitan Vancouver region. Cutting, sewing, and filling, pattern-making tests to stitching experiments; the development of Pop Rocks employed an iterative modeling and prototyping process that derived final forms from the material logic of using fabric to contain granular aggregates. This responsive process-based methodology and its results are indicative of design operations that are increasingly relevant in the context of decreasing resources. Radically recycled architectures, such as Pop Rocks, mark a departure from traditional top-down form-heavy design methods towards a contingent, emergent, and tactical design ethos. This might be described as a new form of pragmatism that is not only ethically enticing but also promises new aesthetic, formal, social, and political frontiers. The soft suppleness of waste finds its avatar in built environments that challenge the dominance of the hardness in cities and its associated behavioral norms. The presence of soft forms at the core of the city extends the typical range of active and passive social activities fostering unexpected social encounters and new perspectives on the city.

我们的设计渗入了战略性的资源使用理念，该委托的暂时性突出了有效利用资源的必要性。该项目完全是由温哥华大都市区域消费后以及加工后的废料建成。进行了切断、缝合以及模型制作试验；运用迭代建模和原型法开发跳跳糖广场，从已使用建筑物的材料逻辑中得出最终形态，以包含颗粒状集料。这种基于流程的方法及其结果在设计过程中可以节省资源。从本质来讲，跳跳糖广场这种再利用建筑与传统的面向偶然、紧急和战略设计理念的设计方法相背离。可以将之描述为一种新的实用主义手法，其理论十分迷人，另外还能呈现出一种美观的、正式的、社交的以及政治的面貌。柔软的废弃物在建成的环境下找到了可用之处，并对城市及其相关行为规范中普遍存在的硬性材料发起了挑战。这些位于市中心的柔软结构扩展了其在各种主动及被动社会活动中的应用范围，促进了人们之间的会面与交流，并产生了新的视野。

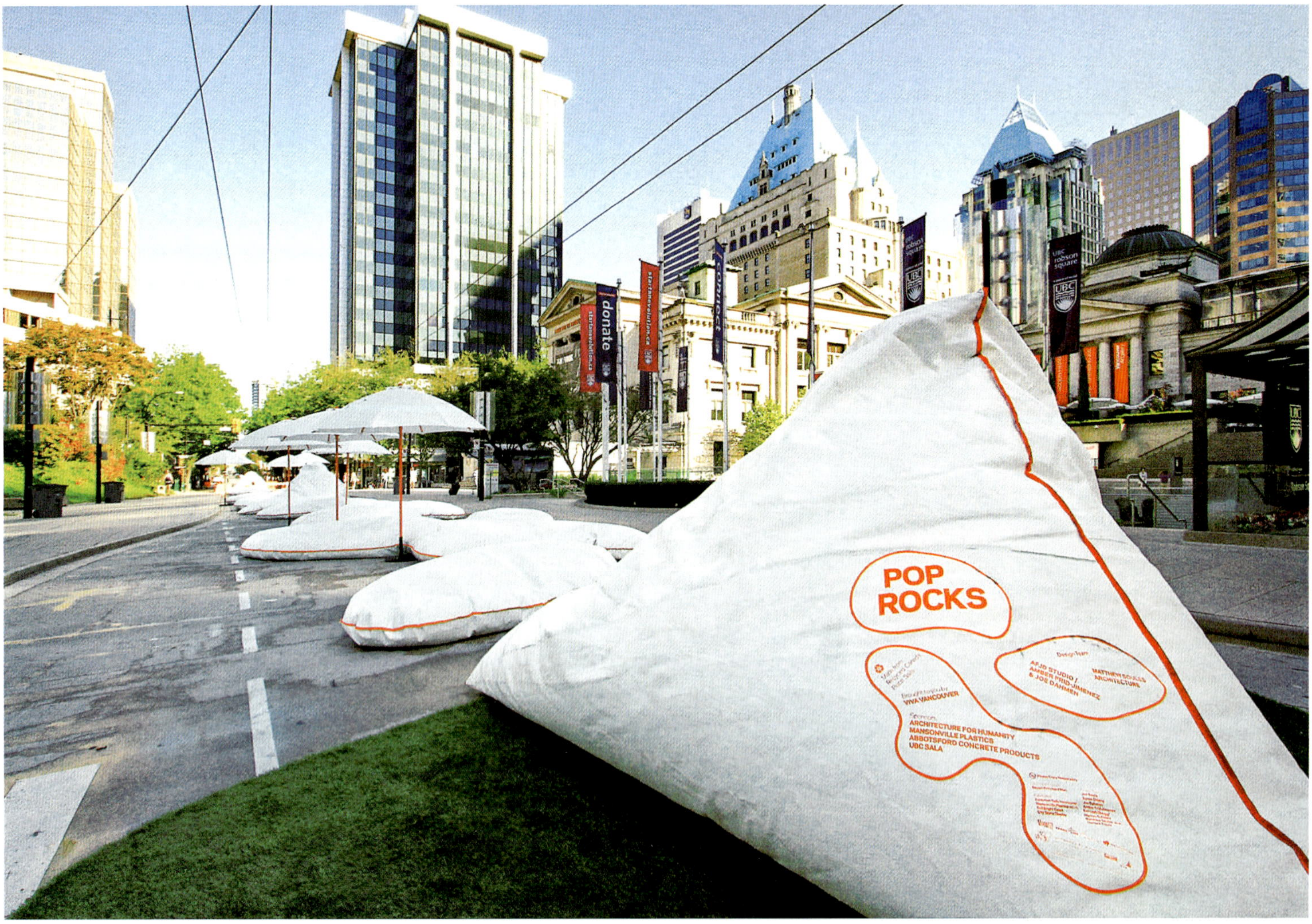
POP
ROCKS
Design Team
AFJD STUDIO /
AMBER FRID-JIMENEZ
& JOE DAHMEN
MATTHEW SOULES
ARCHITECTURE
VIVA VANCOUVER
ARCHITECTURE FOR HUMANITY
MANSONVILLE PLASTICS
ABBOTSFORD CONCRETE PRODUCTS
UBC SALA
donate

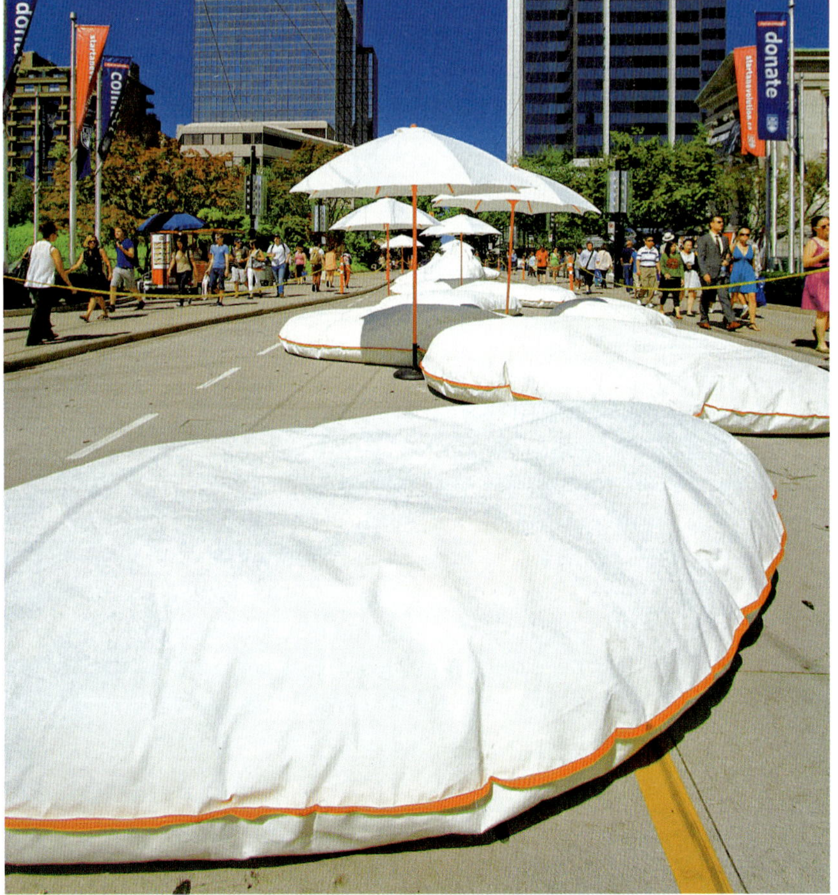

donate

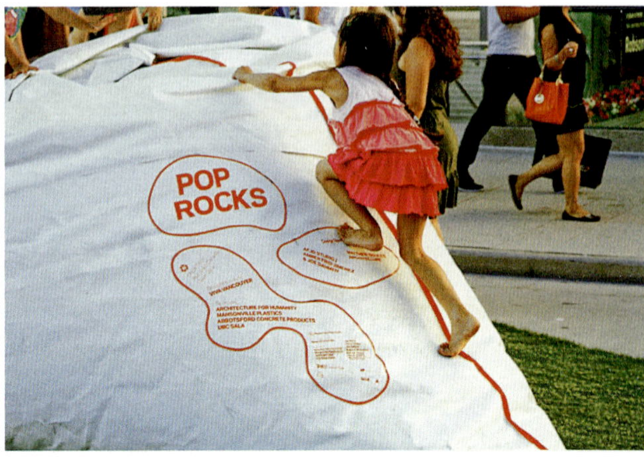

POP
ROCKS

光宝电子总部广场

Major Taiwanese electronics company Plaza

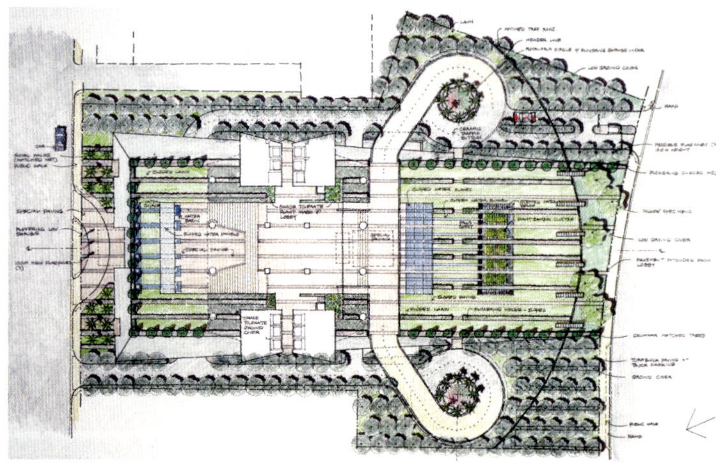

LOCATION:
TAIWAN，CHINA
ARCHITECT:
ARTECH-INC
LANDSCAPE ARCHITECT:
TZUI-YI TSAI, INNERSCAPES DESIGNS

项目地点：
中国台湾
建筑师：
ARTECH-INC
景观设计师：
Tzui-Yi Tsai, Innerscapes Designs

Lite-On, a major Taiwanese electronics company, chose the site for their new corporate headquarters with the "Electronics Center" of Taipei overlooking the Gee Long River. The owner retained the architects and landscape architects to work together to achieve three key objectives:

光宝集团是台湾主要的一家电子公司，选择了这块地方作为他们新的公司总部，位于眺望着基隆河的台北电子中心。公司所有者聘用建筑师和景观建筑师一起工作以实现三个主要目标：

1. Design an energy efficient, sustainable "green" building.
2. Integrate the building and site with the landscape as an important component of being "green".
3. Provide secure and functional outdoor space for passive use by employees and visitors.

1. 设计一个高效节能、可持续的绿色建筑。
2. 把建筑和场地融为一体，让景观成为绿色概念的重要组成部分。
3. 为员工和游客提供安全的功能性户外空间。

The overall concept was a 25-story slender tower rising above a sloped landscape podium that covered much of the site. The podium built over four levels of below grade parking sloped toward the river on one side and toward the urban center on the other side. This configuration maximizes views from the tower and for users of the landscape gardens. The podium provided security, view gardens and a green roof, retaining storm water, storing it for irrigation, and providing insulation for the public spaces below.

整体的规划是在覆盖大部分场地、突起倾斜的景观底座上修建一栋25层高的修长的高楼。底座之下是四层高的地下停车场，一边向河的方向倾斜，另一边朝向市中心。这样的布局使景观花园的使用者能从高楼上最大化地欣赏景色。底座提供了保安功能、观赏花园和一个绿色屋顶，屋顶能盛装用于浇灌的雨水，也为下面的公共空间提供了隔离保护。

The podium at the rear of the site slopes one story to the street where a major vehicular entry provides access into the recessed atrium court. The larger podium on the riverside slopes two stories down to the boulevard. A second vehicular drop off is on this podium at the front building face. As the garden slopes away from the tower stepped water features create visual and audio interest. The landscape features take into account the limitations on the soil depth, weight, etc., imposed by the structures below.

场地后面的底座向大街倾斜一层楼高，在那儿一个主要的车辆入口直通到里面的中庭。河边一侧的底座向下倾斜两层楼高一直到林荫大道。另外一个车辆落客区在高楼前面的底座上。因为花园从高楼向下倾斜，阶梯型水景营造出视觉和听觉上的享受。各个景观设施都考虑到了下面的建筑结构对土壤深度、重量等方面的限制。

The view corridor of the long sloped podium is emphasized by linear planting beds, lawn panels, and granite walkways. Garden spaces are created to allow for informal seating and outdoor gatherings. Halfway down the garden, a light well opens to a courtyard below that is viewed from the cafeteria at the lower level. Stepped fountains in the courtyard and a grove of Madagascar almond (Terminalia mantaly) can be viewed from the cafeteria or from the pedestrian bridges on the podium above.

底座倾斜的观景长廊主要是线性的植物带、一些块状草坪和花岗岩步行道。花园空间可以坐着休息和户外聚会。往下方花园走到一半的地方有一个天井，下面是一个庭院，坐在下层的餐厅可以观赏它。庭院里阶梯型的喷泉和小叶榄仁树林可以从餐厅或从上方底座的人行天桥欣赏到。

The two podium gardens are flanked on the street below by orchards of Camphor trees and Golden Rain trees also serving as street trees. The generous open space around the tower gives it distinction in a dense urban context, as well as playing a major role in accomplishing the vision of the owner as a total "green" development. Conceived before LEED accreditation, this was the first green roof envisioned and built by a private party in Taipei.

在下面的街道边是两个底座花园，园边上栽植着樟树和栾树的小树林，也起到了街树的作用。高楼四周开阔的开放空间使它在稠密的城市背景里格外突出，也在实现业主的全面绿色发展的设想中发挥了重要作用。它的构想先于绿色能源与环境设计体系认证，是台北市第一个由私人团体设想与建造的绿色屋顶。

广东省深圳市中科研发园广场

Technology Park Landscape Design, ShenZhen GuangDong

LANDSCAPE ARCHITECT:
Guangzhou Turen Landscape Planning CO.,LTD
CHIEF DESIGNER:
PangWei
AREA :
22705.1m^2

设计单位:
广州土人景观顾问有限公司
首席设计师:
庞伟
面积:
22705.1m^2

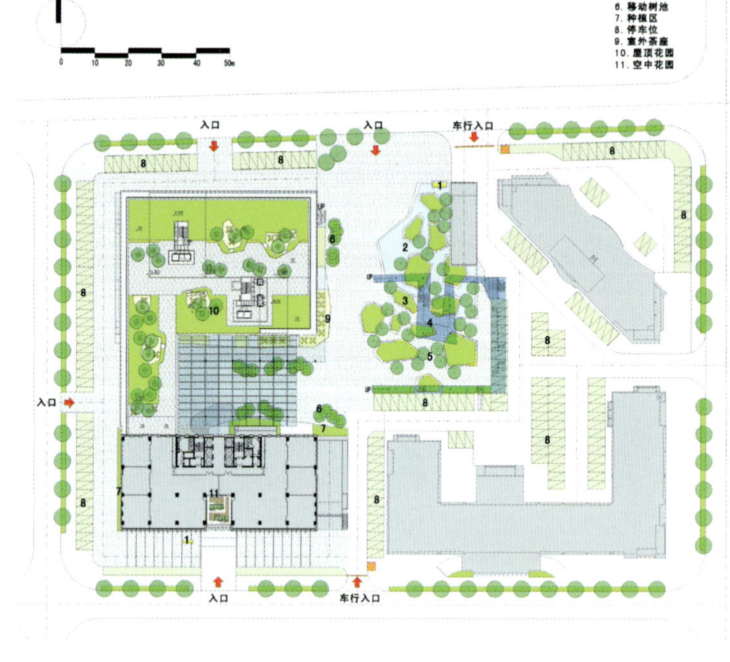

The site without open space being one of city blocks, how it becomes inventive and attractive public space?

一个不大的项目，普通的城市地块，如何构建实用、有魅力的城市公共空间？

The compression of the ShenZhen technology park project located SouthHill, Shenzhen, in the narrow and disorder space between old office blocks and both with new buildings. With complicated contours, abnormal field identification, messed up parking system, the site being very common city space in China. It is aimed to create practical, positive and charming urban public space.

深圳中科研发园位于深圳南山高新技术区南区，身处新旧夹杂的多层办公楼与高层写字楼包围之中。场地标高流线复杂，归属与形态不明确，周边多家单位停车问题复杂，正是中国常见的混乱无序的普通城市地块。如何营造实用的、积极的、有魅力的城市公共空间，成为本次设计的目的和挑战。

The premises of design generate a bended bridge embedded in unusual-shape water container and planters, becoming a main scene in the agro. The flows of transportation rely on the mainstream path on the bridge, articulating pieces of space. Under the bridge, sunshine is brushed up into shadows, drawing the possibility of discovery. On the bridge, the space is divided into two scales, which related to the function of path way and the rest to be leisure space.

一座曲折错落的景观桥体镶嵌于异形水池与植物种植容器之上，成为主景。围绕桥体组织交通流线，链接离散空间。桥底设置数排光影柱，轻盈宛如波光流动。桥面分割成不同尺度的踏面，分作交通与休憩之用。桥下遮阴，桥上休憩。以形态本身来满足人性需求。

Simultaneously, people might become the part of scene, dramatically watching.

同时，人在行动中滞留，成为风景。看与被看，正是戏剧的张力之美。

过山车广场
Rollercoaster

LOCATION:
Beijing Huangzhuang Vocational School
DESIGN COMPANY:
Interval Architects
AREA:
1200m²
PHOTOGRAPHY:
GU Yunduan
PROJECT ARCHITECT:
Oscar KO, GU Yunduan
LIGHTING CONSULTANT:
MIAO Hailin

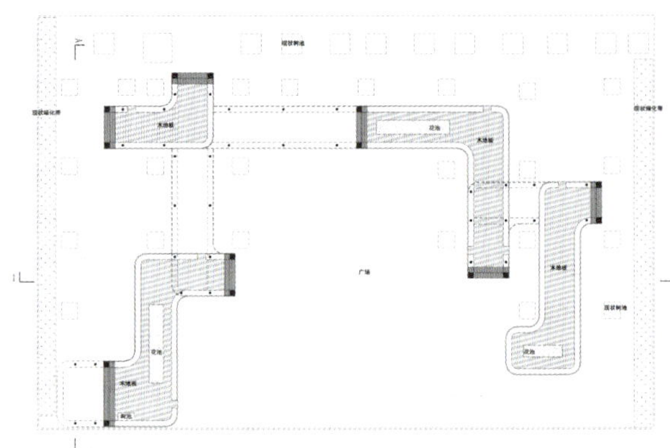

项目地点：
北京黄庄职业高中
设计公司：
空格建筑
面积：
1200m²
主持建筑师：
高亦陶，顾云端
灯光设计顾问：
缪海琳

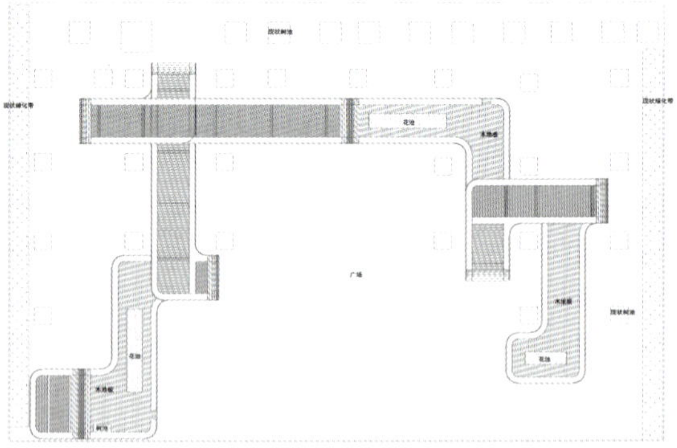

Situated in a tranquil environment of one of the best vocational schools in Beijing, the project aims at providing an iconic image to the institution as well as redefining the use of an existing public space on the central square of the campus.

该项目位于一所宁静的中学校园的活动广场之上，旨在为学校提供一个鲜明的可辨识的形象，并重新定义现有公共空间。

Initially, the client wanted to put on the square a themed sculpture with a monumental effect and scale. A huge pedestal was even already built for the sculpture to put on. However the obvious problem of the square is actually a severe lack of effective public space that would allow students to gather and communicate. What the school really needs is not a monument in the center of the campus, but a humanistic and functional gathering space for students and an event space for school activities.

最开始客户的设想是在广场上安放一个纪念性的主题雕塑，并且已经为该雕塑搭建好了一个高台。然后通过分析，我们认为广场真正缺少的是一个有效的能让学生聚在一起交流的公共空间。学校真正需要的不是校园中心的纪念碑，而是一个集人文与功能为一体的集聚性空间，让学生休憩的同时还能举办一些校园活动。

Therefore, with the intention to create an efficient public space, we proposed a continuous self-folding belt structure that resembles the image of a "roller coaster". The structure can be perceived as a ribbon that folds and bends three-dimensionally to create a series of spaces such as open gardens, shaded pavilions and exhibition corridors. As the roof beams of the pavilion bend downwards and become wall screens and benches, the structure expresses a formal continuity well combined with different functions. The pavilion snakes around the existing trees on the site so they are well-preserved and maximally utilized for shading.

因此，一个高效的公共空间的方案渐渐浮出水面。我们提出了一个连续弯折类似"过山车"的带状形象，通过三维折叠，创造了一系列与环境融合的空间，包括开放式的花园、阴影展馆和展览走廊等。整个建筑的弯曲形态都最大限度地考虑了保留广场现有的树木以及利用现有树冠投射下的阴影，使建筑建成的伊始就具有成熟的使用形态。

Therefore, with the intention to create an efficient public space, we proposed a continuous self-folding belt structure that resembles the image of a "roller coaster". The structure can be perceived as a ribbon that folds and bends three-dimensionally to create a series of spaces such as open gardens, shaded pavilions and exhibition corridors. As the roof beams of the pavilion bend downwards and become wall screens and benches, the structure expresses a formal continuity well combined with different functions. The pavilion snakes around the existing trees on the site so they are well-preserved and maximally utilized for shading.

因此，一个高效的公共空间的方案渐渐浮出水面。我们提出了一个连续弯折类似“过山车”的带状形象，通过三维折叠，创造了一系列与环境融合的空间，包括开放式的花园、阴影展馆和展览走廊等。整个建筑的弯曲形态都最大限度地考虑了保留广场现有的树木以及利用现有树冠投射下的阴影，使建筑建成的伊始就具有成熟的使用形态。

哈格威尔地产广场
Hageveld Estate

LOCATION:
Heemstede, the Netherlands
AREA:
730 square meters (pond),
14 hectares (estate)
DESIGNERS:
Berrie van Elderen, Marike Oudijk
PHOTOGRAPHER:
Pieter Kers

项目地点：
荷兰 Heemstede
面积：
730 平方米（池塘），14 公顷（土地）
设计师：
Berrie van Elderen, Marike Oudijk
摄影师：
Pieter Kers

Hosper drew up the urban and landscape master plan, a design for the land of the Hageveld estate and the (detailed) design of the ornamental lake on top of the underground car park.

Hosper 为 Hageveld 庄园制定了城市与景观的总体规划，以及地下停车场顶部的观赏湖的（细节）设计。

The estate had previously been a seminary, and part had been used as a secondary school. Although Hageveld still had allure, the vegetation has been neglected and the Voorhuis (front section) was vacant. In 2002 the new owner wished to transform the Voorhuis into a block of (luxury) apartments. At the same time the Hageveld Atheneum College wished to expand the Achterhuis (back section).

此房地产项目此前是神学院，部分建筑还曾经是一所中学的教学楼。虽然 Hageveld 仍然有些吸引力，但植被已无人照管，Voorhuis（前面部分）也是空置的。2002 年，新业主希望把它转化成（豪华）公寓区。而同时，Hageveld 雅典学院希望扩大 Achterhuis（后院部分）。

The spatial design of the estate is based on three concentric shells, each with an individual character. The main building forms the core, which is surrounded by the lushly-planted estate and, outside the estate, farmland. The two outer shells function as a (landscape) park.

房地产的空间设计基于三个有各自特点的同心圆壳。四周植被茂盛的主体建筑构成了其核心，其外围则是农田。外侧的两个圆壳起到了（景观）公园的作用。

HAGEVELD

嘉瑞轻纺广场
Hageveld Estate

DEVELOPERS:
Jiarui Textile Co., Ltd., Zhejiang
LOCATION:
Hangzhou
COVERING AN AREA:
38712M²

开发商：
浙江嘉瑞轻纺有限公司
项目地点：
杭州
面积：
38712 平方米

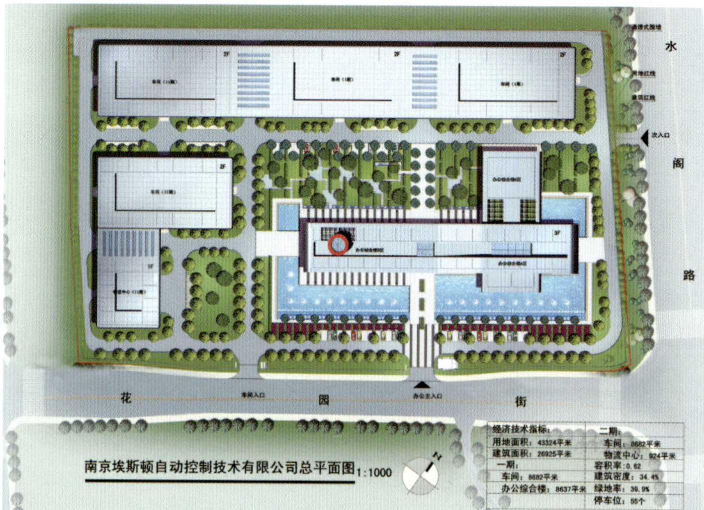

Four-Season Patio: It is not only a square with leisure breath, but also an outdoor living room that is full of green. For the purpose of working and guest reception of the office building, the concept of leisure office and guest reception is brought from inside to outside, and turn "four-season patio" into a scenic resting and reception courtyard. The patio creates a pleasant gathering place that is rich in modern atmosphere. The flexible and changeable space, the green landscape with multiple layers and careful arrangement, harmonize with the multiple functions of the space. Solar terms of four seasons are the keynote of the square, while the change of the four seasons is displayed through the carefully furnished waterscape design, flowers and wood. Water is the important part of this square in respect of layout; different shapes of water symbolize the features of different solar terms, creating an eyeball-attracting effect at the entrance, and producing tranquil and leisure atmosphere for the gathering place.

四季中庭：位于嘉瑞轻纺办公楼前院，是个具休闲气息的广场，又是绿意盎然的户外居庭。从办公楼的办公与接待宾客的本质出发，把休闲办公和迎宾的概念自传统的室内推展至室外的庭园空间，让【四季中庭】成为休憩、迎宾的景观庭院。中庭以富现代感的空间基调来构建一个宜人的聚会场所。场所内有充满灵活性和可变性的空间，层次丰富、布局仔细的绿化景观配置，配合多功能的空间特性。广场以四季节气为设计的重心，而四季的演化则以精心配套和安排的水景设计和花木所呈现。水是这广场布局上的重点部分。它的各种形状象征了不同节气变化的特色，为入口部分提供光彩引人的效果，同时为聚会场所营造憩静、闲息的空间感。

法院广场
Court Square

GENERAL CONTRACTOR:
AJ MARTINI
ARCHITECT:
John Cunningham Architects, Inc.
ARCHITECT (LOBBY):
Office da Landscape
CONTRACTOR:
Emanouil Inc.
STRUCTURAL ENGINEER:
DM Berg Consultants

总承包商：
AJ Martini
建筑师：
John Cunningham Architects, Inc.
建筑师（大堂）：
Office da Landscape
景观承包商：
Emanouil Inc.
结构工程师：
DM Berg Consultants

Wonderful, playful, original, and it will only get better with time. The landscape architect created spatial variety and has a wonderful sense of color. A little jewel.

这个项目精彩纷呈、有趣而新颖，它只会随着时间的推移而变得更美好。景观设计师创建了富于变化的空间，并有极佳的色彩感。宛如一颗小小的宝石。

林肯大街 1111 号广场

Lincoln Road 1111 a Miami

LOCATION:
US, MIAMI
ARCHITECT:
Herzog & de meuron
LANDSCAPE ARCHITECT:
Raymad Jungles Inc

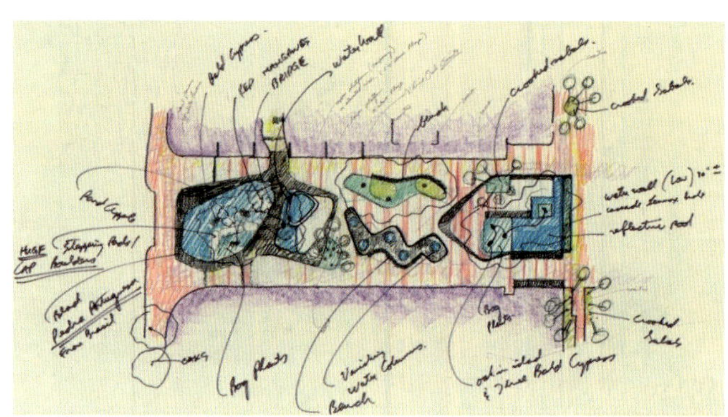

项目地点：
美国 迈阿密
建筑师：
Herzog & de meuron
景观设计师：
Raymad Jungles Inc

本案的景观设计师 Raymond Jungles 是美国景观设计理事会会员，受雇成为这个景观项目的首席设计师，他和 Herzog&deMeuron 建筑事务所一起为林肯大街最西部这个长 355 英尺、宽 100 英尺的街区进行重新设计。林肯大街 1111 号的设计追求大胆、简单以及永恒经典的设计风格。委托人要求设计师 Raymond 的设计要兼容 Morris Lapidus 最初对户外热带购物环境和林肯大街美食街设计的构思设计。设计师力图以佛罗里达州当地的景观为动物提供栖息地和庇护所，同时也可以对这里的行人进行一些教育启发。最终设计师对这里的设计目标就是为游客和居民建造一个健康、繁荣、永恒的城市景观。

玛莎塔德医院（中庭）广场
Maasstad Hospital

LOCATION:
Rotterdam, The Netherlands
SITE AREA:
±4000 m^2
DESIGNER:
Stijlgroep landscape and urban design
PHOTOGRAPHER(S):
Stijlgroep landscape and urban design

项目地点：
荷兰鹿特丹
占地面积：
约 4000 平方米
设计师：
Stijlgroep landscape and urban design
摄影师：
Stijlgroep landscape and urban design

The large new hospital building with its sober and traditional comb structure links together five courtyards which bring daylight into the surrounding rooms. The patios are located along the central internal corridor, which connects the entire building from one end to the other.

稳重且传统的梳状结构的新医院大楼与五个庭院连接在一起，给周围的房间带来了明亮的阳光。中庭位于沿中央内部走廊而设置的庭院，走廊从一端到另一端连接着整个建筑。

The design team of Stijlgroep decided to provide each patio of the three building sections with a different topic. Based on the concept of contrast and the idea that the patios will function as orientation points for patients/ staff and visitors our design team came up with a unique design.

Stijlgroep 设计团队决定为每个庭院的三个组成部分都提供不同的主题。基于对比的概念以及庭院将为患者 / 工作人员和游客提供定位点的想法，我们的设计团队提出了一个独具一格的设计方案。

Every patio has a fairly minimalistic layout being complemented with different themes of plantation. The main focus in all five patios rests on the very special lighting objects.

每个庭院都有相当简约的布局，补充以不同主题的种植园。所有五个庭院的主要焦点都在于在非常特殊的照明对象。

In the first two patios massive rectangular lighting columns emphasise the minimalistic design of the patio whereas oversized seating elements, which are lit during evening and at night time, are the main element of the central patio. Those seating elements dare users to sit under the soft shade of the huge honey locust trees which were planted. In the last two patios gigantic bamboo-like lighting elements are placed in a modern interpretation of a bamboo garden.

在前两个庭院中，巨大的长方照明柱强调了庭院的简约设计，而在傍晚和夜间点亮的超大座位，是中央庭院的主要元素。这些座位元素使用户可以坐在种植的巨大皂角树的树荫下乘凉。在后两个庭院中，如同巨大的竹子的照明元素被放置在一个有现代感的竹园里。

What seems very special already at daytime is becoming even more special when night falls and the lighting elements are illuminated by multi-colour LED lighting.

在白天已经似乎很特殊的由多色 LED 提供的照明元素，当夜幕降临的时候就变得更特殊了。

The patios themselves create a quiet and secure oasis for patients, visitors and staff. Additional advantage of these green patios is the positive affect on recuperating patients. Green spaces have simply a very positive influence on people referring to the concept of a healing environment.

庭院本身为病人、访客和工作人员创造了一个安静且安全的绿洲。这些绿色庭院的额外优势是对休养病人具有积极的影响。根据愈合环境的概念，绿色空间会对人们产生非常积极的影响。

Stijlgroep Landscape and Urban Design was not only commissioned with the design phase, but also with the detail construction and the on site management and supervision. Due to our strong expertise the project has been delivered on time and within budget.

Stijlgroep 景观和城市设计事务所不仅负责设计阶段，还负责具体的建设、现场管理和监督。由于我们强大的专业能力，该项目已在规定时间和预算范围内交付。

美的总部大楼广场
Landscape Design for Midea Group Headquarter

LOCATION:
Foshan, Guangdong
SITE AREA:
23,000 m^2
CLIENT:
Guangdong Midea Co., Ltd.
LANDSCAPE ARCHTENT:
Guangzhou Turen Landscape Planning Co.,LTD
CHIEF DESIGNER:
Pangwei
PROJECT LEADER:
Zhangjian

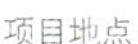

项目地点：
广东省佛山市
占地面积：
23 000 平方米
完成时间：
2009 年
委托人：
广东省美的电器股份有限公司
景观设计：
广州土人景观顾问有限公司
首席设计师：
庞伟
项目负责人：
张健

The landscape design of Midea Headquarter Building responds to the southern landscape of China - mulberry fish pond by means of modern landscape language, returning to rural scenic forms and native aesthetic conception at this highly-modernized times. The crisscrossing trestle and road divide the land into many geometrical bodies of different sizes and forms – some parts sink and are turned into waterscape; some parts rises and are turned into the small hills that are used to plant rural trees; some parts are built into small square (courtyard), or the basement areaway of the basement. The land is dotted with modern scenic structures built with rural materials, running through and extending the regional landscape. The combination of form and rural materials solve the visual problem of the areaway of some raised basement. The actual functional matrixes such as trestle, road and waterscape draw the outline of the cell texture of mulberry fish fond, making people feel the lively and rich functional connections of different textures; in addition, the texture of the intimate and mild fish pond brings people back to the past, giving a sense of belongs of the land.

美的总部大楼景观设计通过现代景观语言回应中国岭南大地景观“桑基鱼塘”，在高速城市化的当下回归乡土景观形式与本土美感意境。阡陌交通的栈桥和道路将用地分割成大小不等、形态各异的几何体——或下沉为水景，或上浮为种植乡土林木的小丘，或成为区域小广场（庭院），或是地下室采光天井。并在其上点缀以乡土材料建造的现代景观构筑，以形态和乡土材料组合解决高起的若干地下室采光天井的视觉问题，贯穿、延续地域景观。用栈桥、道路、水景与庭院等实际功能体块勾勒出“桑基鱼塘”的网状肌理，让人体验到的不仅是肌理间生动丰富的功能联系，还有亲切舒缓的鱼塘肌理带给人的仿佛当年人对土地的归属感。

Waterscape consists of ecological wetland and the thin water above the basement areaway, and the purpose is neither to re-present the different forms of water, nor to realize the water-related activities, but to show the respect to the relationship between regional culture, living and natural environment. The wetland and the thin water above the basement areaway are designed into a part of the rainwater and sewage water collection system. The rain water on the roof and the open area is collected, treated and stored in the scenic water pond. The reclaimed water and sewage water are completely recovered and reused for planting, irrigating and replenishing the water in the scenic water pond through the biodegradation of ecological wetland. Drinking water is not used for scenic water. Waterscape design also interprets the choice we made in the face of current environment conditions – how to continue and exist.

水景在场地中被分作生态湿地以及地下室采光天井上的薄水之用，其重点不在于再现水景的不同形式，也不在于水景带来的若干亲水活动，而是在于对区域文化、生活及当地自然环境关系的尊重。生态湿地以及地下室采光天井上的薄水皆设计为雨水、污水收集处理系统的一部分。把屋面和露天雨水收集、处理、蓄积在景观水池之中，将产生的中水和污水全部回收，通过生态湿地进行生物降解处理，回用于绿化灌溉和补充景观水池水量，不使用饮用水作为景观用水。水景设计也解释了我们面对今天环境现状所做出的选择——如何延续和存在。

明尼苏达大学德卢斯广场
University of Minnesota Duluth

LANDSCAPE ARCHITECT:
oslund.and.assoc. - Thomas Oslund FASLA, FAAR; Misa Inoue, RLA
ARCHITECT OF RECORD FOR THE BUILDING:
Stanius Johnson Architects, Inc.
DESIGN ARCHITECT FOR THE BUILDING:
Ross Barney + Jankowski, Inc.
STRUCTURAL ENGINEER:
Meyer, Borgman and Johnson, Inc.
CIVIL ENGINEER:
MSA Professional Services
MECHANICAL & ELECTRICAL ENGINEERS:
Affiliated Engineers
GENERAL CONTRACTOR FOR THE BUILDING:
M.A. Mortenson
GENERAL CONTRACTOR FOR THE SITE WORK:
Max Gray Construction, Inc.
COST CONSULTANTS:
Oscar J. Boldt Construction Co.

景观设计师：
oslund.and.assoc. – Thomas Oslund FASLA, FAAR; Misa Inoue, RLA
建筑师：
Stanius Johnson Architects, Inc.
建筑设计师：
Ross Barney + Jankowski, Inc.
结构工程师：
Meyer, Borgman and Johnson, Inc.
土木工程师：
MSA Professional Services
机电工程师：
Affiliated Engineers
建筑总承包商：
M.A. Mortenson
现场总承包商：
Max Gray Construction, Inc.
成本顾问：
Oscar J. Boldt Construction Co.

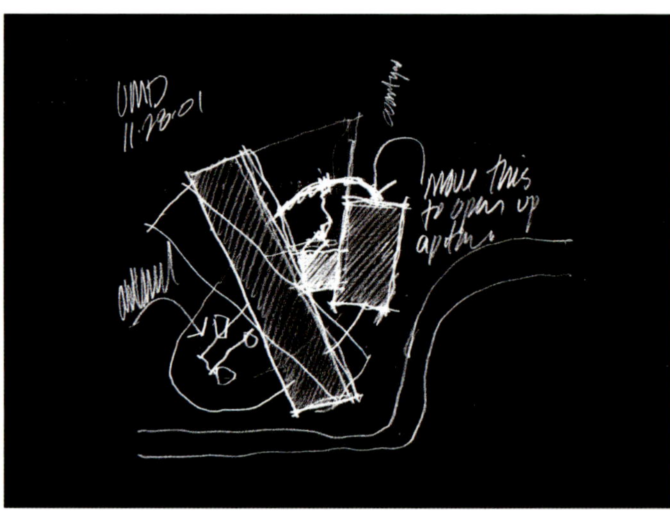

The site plan for the new Swenson Science Building takes its cues from the concept of "Science on Display." Clad in steel, and surrounded by native plantings – resources unique to northern Minnesota – the courtyard features an experimental wetland garden, with an emphasis on the cultivation of wild rice. Chosen for its symbolism and importance to the Native American population of northern Minnesota, the garden is an outdoor laboratory for UMD students and faculty.

新斯文森科学楼的设计思路源自＂展示科学＂的概念。大楼由钢铁覆盖，周围生长着很多明尼苏达北部特有的植物品种，校园里有一个实验用湿地花园，主要用来耕种野稻。 因为其对明尼苏达北部的美印第安人的象征意义和重要性，这个花园是明尼苏达大学德卢斯校区学生和教员的户外实验用地。

"Science on Display" embraces the idea that "seeing" is critical to "understanding", thus the new Swenson Science Building (SSB) at University of Minnesota, Duluth, incorporates an atrium and glass-walled classrooms, as well as interactive outdoor learning & contemplative spaces. Clad in steel, slate, and brick, the SSB will be one of three landmark buildings that act as gateways onto campus. The building operates as a bridge over one of the main campus arterials, tying the new space to existing science buildings. The act of adding a sense of transparency to the design of both the interior and exterior rooms allows activities normally conducted in opacity to be revealed. This helps those passing through the building to understand its function, or at least have their interests inspired by what they see.

"展示科学"包含这样的观点："看见"对"理解"至关重要，于是明尼苏达大学德卢斯校区的新斯文森科学楼有一个中庭、玻璃墙教室，还有互动的室外学习和思考的空间。科学楼由钢铁、石板和砖头构建，将作为校园入口的三大地标式建筑之一。该建筑还起到了桥梁作用，横跨校园的一条干路，把新的空间和已有的科学建筑相连。内外房间的设计增加了一种透明感，将通常在不透明状态下进行的活动暴露出来。这有助于那些穿过楼的人理解它的用途，或者至少让他们对所看见的东西产生兴趣。

唐恩都乐广场—地平线花园广场

Dunkin Donuts Plaza — Horizon Garden

PROJECT LOCATION:
Providence, Rhode Island
SIZE:
6,000 sq ft
PHOTOGRAPHER:
Mark LaRosa

项目地点：
罗德岛普罗维登斯
面积：
6000 平方英尺
摄影师：
Mark LaRosa

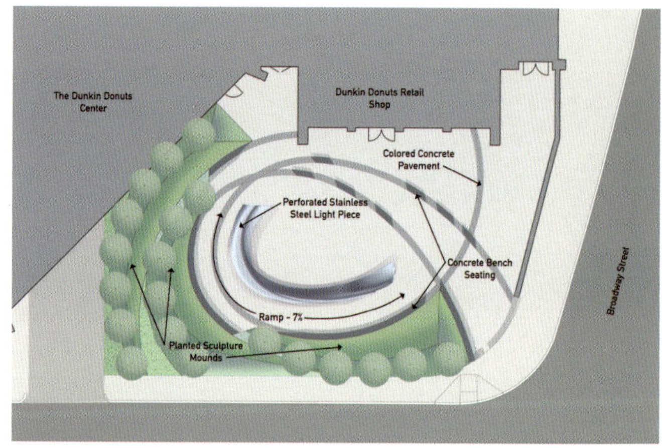

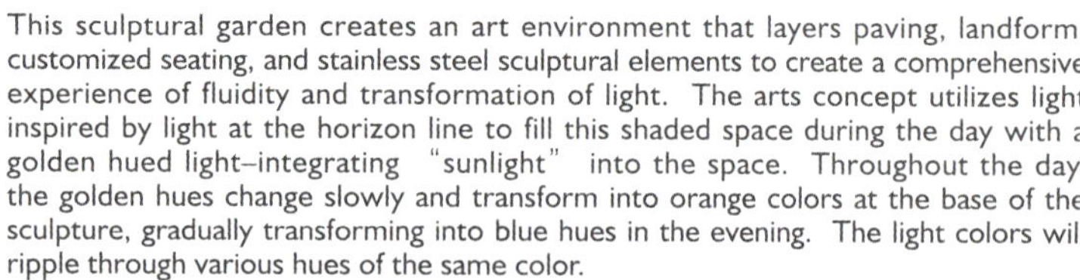

This sculptural garden creates an art environment that layers paving, landform, customized seating, and stainless steel sculptural elements to create a comprehensive experience of fluidity and transformation of light. The arts concept utilizes light inspired by light at the horizon line to fill this shaded space during the day with a golden hued light–integrating "sunlight" into the space. Throughout the day, the golden hues change slowly and transform into orange colors at the base of the sculpture, gradually transforming into blue hues in the evening. The light colors will ripple through various hues of the same color.

这个雕塑花园把铺设的地面、地形、特制的座椅，和不锈钢雕塑层次有序地布置起来，营造出艺术的氛围，充分体现了光的流动与变幻。这个艺术概念从地平线的光获得灵感，白天使用金黄色调的光来充满这个背阴的空间，就像把阳光引入这个空间。整个白天金黄色调慢慢改变，在雕塑底端变成橙色，在夜晚渐渐变成蓝色调。同一颜色的光通过不同的色调泛起涟漪。

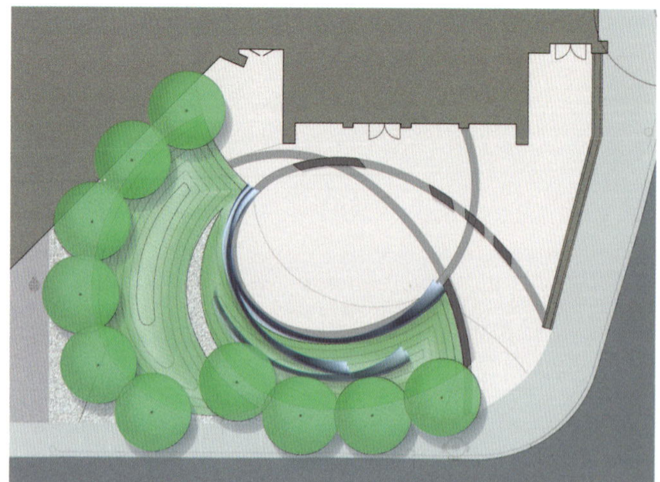

The stainless steel sculptural element frames the various spaces and activities in this art garden. The sculptural surface is a folded and layered perforated stainless steel skin that allows for light and color to emanate through the textured surfaces of the piece. During the day, golden hues varying from bright yellow to an orange yellow emanate from the translucent surfaces brightening the space and creating a warm focal point of "sun on horizon". At night, the sculpture becomes a transforming piece of blue light and fluidity within the plaza. The sculpture encourages participants to move around and engage with the various garden spaces framed by the paving patterns, landforms and plant materials.

在这个艺术花园里，不锈钢雕塑围起不同的活动空间。雕塑的外表是一块层叠的带网孔的不锈钢板，可以让光色透过其织纹状的表面。在白天，金黄色调从明黄变为橙黄，从半透明的表面散发出去，照亮了空间，创造出引人注目的太阳照耀地平线的温暖景致。在夜晚，雕塑在广场内变成了一个蓝光的流线体。雕塑吸引着人们来到这个由不同图案的地面、地形和植物构成的不同空间里活动。

乌得勒支广场
Inktpot Utrecht

LOCATION:
Utrecht Netherlands
SIZE:
3000m^2

项目地点：
荷兰 乌得勒支
面积：
3000 平方米

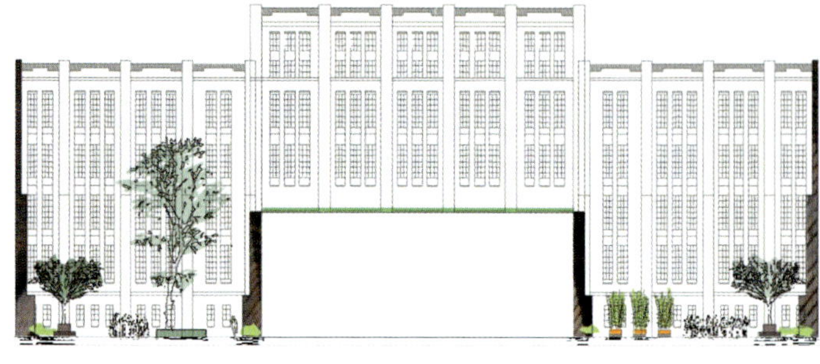

Nothing in the present, utilitarian-designed patio reflects a design for the building's function as logistic centre of the railway network or the significance of the railway organisation. In the plan developed, the seeming contradictions and limitations are utilized to create a patio that speaks to the imagination. The design sets the scene for various uses, both the stage and its sets and the world behind the scenes. The current "utilitarian" use is supplemented with more of a "use for enjoyment", which is nonetheless just as functional, but of another order: sitting in the summer sun, getting outside together and having a break meeting while enjoying the green. The third function is that of "green for the eye". The roof of the building in the inner courtyard can fulfil this function for the users of the upper floor offices adjacent to the interior space. It becomes a green roof, utilizing the very limited load-bearing capacity of the building.

现代、实用的中庭让人丝毫感觉不到这是一座铁路物流中心建筑以及铁路组织的重要性。规划采用表面上的矛盾与限制，打造出了一个充满想象力的中庭。设计实现了各种功能，包括舞台及其背景，以及景观后面的空间。实用性与娱乐性相得益彰，不仅功能性强，而且还可以用作其他用途：坐在夏日和煦的阳光下，在享受绿意盎然的环境的同时，来一次露天聚会或小憩，多么惬意！第三个功能是"为眼睛提供一次绿色盛宴"。对于室内空间相邻的上层办公室的人来说，内庭建筑的屋顶可以实现这一功能；建筑非常有限的承重能力得以充分利用，打造出宜人的绿色屋顶。

The installation of the sculpture was completed during the fall of 2008 and was enhanced with the construction of a new garden grove during the fall of 2009. This garden provided the sculpture with the appropriate spatial and textural setting as originally intended. The sculpture was conceived as a place for quiet reflection and engagement along the edges of the water troughs. These braided troughs provide a subtle babbling sound and cooling effect for the garden.

雕塑是在 2008 年夏天安置的，在 2009 年夏修建了一片小树林，更加衬托出雕塑的美。正如最初打算的那样，花园在空间和构造上给雕塑提供了一个合适的背景。雕塑被设想成一处沿着水槽边缘静思和交流的地方。这些编织在一起的水槽发出潺潺的水声，给公园凉爽的感觉。

ARVEST
Arkansas Department
of Career Education
REHABILITATION SERVICES DIVISION

暂停庭院和山丘草坪广场

Pause Court and The Lawn Hill

LANDSCAPE:
Architect TROP Co. Ltd(TROP : terrains+open space)
WORK TEAM:
Pok Kobkongsanti(Director), Paisit Viratigul, Chatchawan Banjongsiri
LOCATION:
Pattaya, Thailand
AREA:
2,400m^2
CLIENT:
Sansiri PLC
PHOTOGRAPHER:
PirakAnurakyawachon

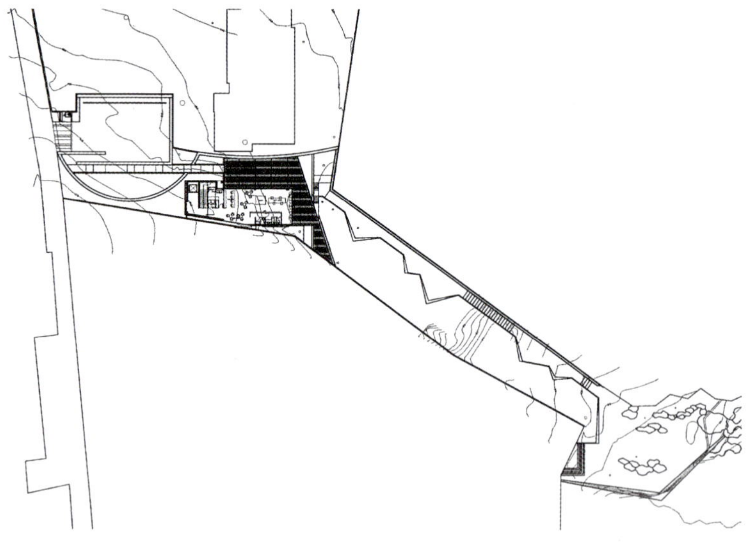

景观设计师：
特罗普（TROP）建筑设计公司（特罗普：地形十开放空间）
工作团队：
Pok Kobkongsanti(Director), Paisit Viratigul, Chatchawan Banjongsiri
项目地点：
泰国芭堤雅
景观总面积：
2400m^2
委托人：
Sansiri PLC
摄影师：
PirakAnurakyawachon

Located in Pattaya, Thailand's perfect destination for low-budget travelers, Baan Plai Haad will be one of the highest priced residential projects in town. Basically our goal is to impress all potential buyers when they visit the site. However, compared to other beachside residential projects nearby, our site is quite a hard thing to sell.

景观位于泰国芭堤雅，是适合低预算旅客的旅游胜地，它将成为市区最贵的住宅项目之一。大体上，我们的目标是在所有潜在买家到访时给他们留下好印象。但是，和附近海滩边其他的住宅项目比较，我们的项目在出售方面特别困难。

The site consists of 2 different characteristic terrains. The First and bigger part is quite flat, locating on top of a small hill, about 12 meters above the beach. The second part is a long, narrow and steep "Tube-like" space that connects the flat land and the beach together.

这个位置由 2 块不同的、有特征的地势构成。第一块，也是较大的一部分，比较平坦，在小山的山顶上，距海滩大概 12 米高。第二部分狭长而陡峭，呈管状，并且和平地、海滩相连。

约翰 E. 贾卡学生运动员学术中心广场
John E. Jaqua Academic Center for Students Athletes

PROJECT LOCATION:
Eugene, Oregon, USA
AREA:
2 hectares

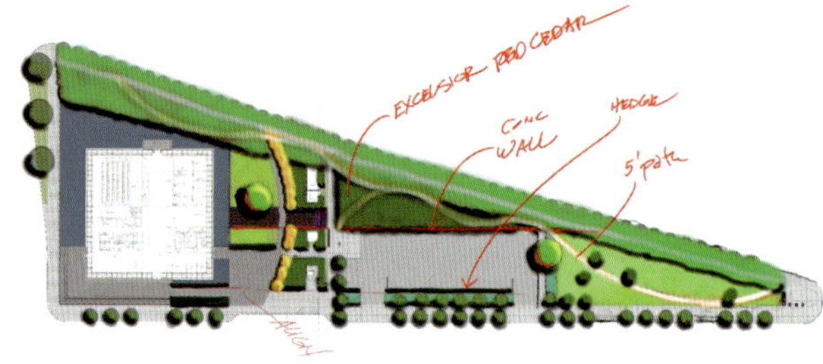

项目地点：
美国俄勒冈州尤金市
面积：
2 公顷

The "Glass Box in a Glass Box on a Black Table of Water with Birches" is the new John E. Jaqua Academic Center at the University of Oregon. It is a pure manifestation of minimalist architecture and landscape architecture, but with a twist. Here, urban ecology is the functional grounds for the living landscape. In this way the yin yang relationship between site and building is clear and determined. You have the very static glass façade that reflects all the seasonal changes and atmosphere of its context while the landscape of mostly native plants, arranged in very orderly patterns, contains all the elements and diversity of a robust "natural" landscape.

这个“有桦树的黑色水台之上玻璃箱中的玻璃箱”就是俄勒冈大学的约翰 E. 贾卡学生运动员学术中心。它是朴素主义建筑和景观建筑的纯粹展现，但有其特色。在这里，城市生态学的反映就是给生活景观提供功能性的那些功能性场地。这样，场地与建筑的阴阳关系是清楚明确的。静态的玻璃外表面反射着所有季节的变化和四周氛围，同时以当地植物为主的景观，排布井然有序，包含了一个通常自然景观应用的所有要素和多样性。

The setting and the building are complementary. With walls of glass, the interior rooms borrow scenery directly from landscape while the landscape uses the glass façade as a reflective backdrop. The strong geometric planes of walkways, plaza' s and reflecting pool stretch beyond themselves, to integrate interior, exterior, and natural spaces into a seamless whole. The raised granite terrace and black table of water evocatively capture the classic modernist conceit by breaking down the membrane between indoors and out, yet the Tables of Water go beyond to mediate between the land, the birches and the sky, to harmonize and funnel each into the buildings interior. As the ultimate experience of the arrival sequence, the raised plinth of the building distills the visitor' s initial landing into one unified gesture of connecting arrival, entry, and an expansive release into landscape.

背景和建筑是互补的。通过玻璃墙，从室内房间可以直接看到外面的风景，景观把玻璃面当成了反射背景。极具几何形状的行走平面、广场和倒影池向远处延伸着，把内部、外部和自然空间连成一个天衣无缝的整体。突起的花岗岩平台和黑色的水池，推倒室内外的隔膜，强烈地表现出经典的现代主义风范，而水池向外延伸承接在天、地和桦树林之间，让它们彼此和谐并把它们反射给室内。建筑的突起底座提供了一种终极的进入体验，把客人的到来成为一种连贯的统一模式：到达，进入，然后是开阔的景观。

Black reflection water surface surrounds the four sides of the building, with linearly arranged birch and local shrub standing at one side. The birch and the surrounding plant communities are the biofiltration system of the project. A beautiful stone entrance square is built at the south side, providing a sunny gathering space for the students, and also serving as the solid base of the building. This "glass box inside the glass box on the black water table with birch" is a main entrance of Oregon University. As a landmark building, it is quite different from traditional campus.

黑色的反射水面环绕在建筑的四面，一侧是直线形的排列优雅的白桦树和当地灌木。桦树以及附近的植物群落是工程的生物过滤系统。一个漂亮的石头入口广场位于建筑的南侧，给学生提供了一个聚集之处，也是建筑的坚固底座。这个“有桦树的黑色水台之上玻璃箱中的玻璃箱”位于俄勒冈大学校园的一个主要入口处，是一个标志性建筑，和传统的校园景观很不同。

约翰内斯堡大学艺术中心广场
University of Johannesburg Arts Centre

LOCATION:
Johannesburg, South Africa
CLIENT:
University of Johannesburg
LANDSCAPE ARCHITECT:
GREENinc
ARCHITECT:
ARC & Mashabane Rose Architects, in joint venture

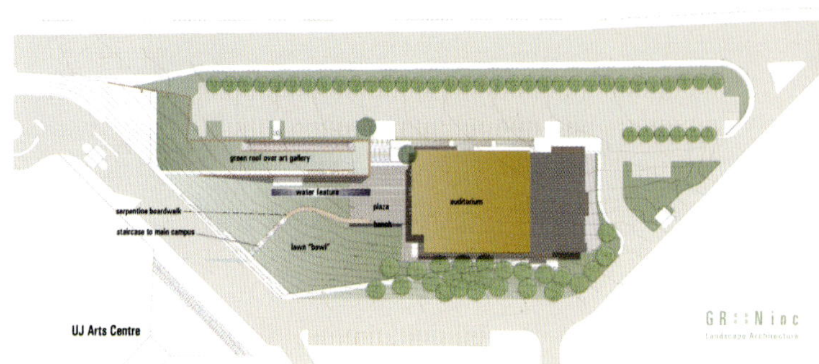

项目地点：
南非 约翰内斯堡
客户：
约翰内斯堡大学
景观设计师：
GREENinc
建筑师：
ARC & Mashabane Rose Architects, in joint venture

Although it does not have formal arts or drama faculties, students of the University of Johannesburg have been staging theatre productions for many years on an amateur basis. The University has been planning to construct a theatre for these productions and an art exhibition venue for almost as long, so the construction of the Arts Centre represents the realisation of a long-standing dream for the university. A wedge-shaped site on the Kingsway boundary of the campus was selected for the scheme. The site is delineated by internal roads on its other three sides, two of which were raised, giving the site a sunken feeling. The University decided to put the design of this most important project out to an architectural competition. GREENinc prepared a landscape design for the scheme entered by Mashabane Rose Architects in association with ARC Architects, and this scheme was selected for construction.

虽然约翰内斯堡大学没有正式的艺术和戏剧学院，但它的学生多年以来一直在进行非专业的戏剧表演。约翰内斯堡大学这么多年以来一直计划为这些艺术作品表演修建一个剧场和一个艺术展览场地，于是大学艺术中心的建设代表实现了一个长久的梦想。位于校园靠 Kingsway 边界的一块楔形的地方被选为项目的场址。场地的其他三面都是校园内部道路，其中的两面突起，给人以场地下陷的感觉。大学决定把这个最重要工程的设计交给一个建筑竞赛。GREENinc 公司准备了一个景观方案设计，其建筑设计部分是 Mashabane Rose 建筑师事务所联合 ARC 建筑师事务所提交的，这个方案被采纳了。

GALLERY
GALERY

Because of the format of the competition, there was not a brief for the landscape architect from the client. In working with the architects on the competition submission, GREENinc's primary goal was to respond to the proposed new buildings and integrate them into the campus.

因为竞赛的模式所限，客户没有向景观设计师做出一个简介。GREENinc 公司在竞赛作品的提交上与建筑师们的主要合作目的是针对那些计划的新建筑，把它们融入校园。

云
Cloud

LOCATION:
Copenhagen, Denmark
AREA:
5,500 m^2
TEAM:
Stig L. Andersson, Hanne Bruun Møller, Signe Hertzum, Malin Blomquist
COLLABORATORS:
Schmidt Hammer Lassen
DESIGN COMPANY:
SLA

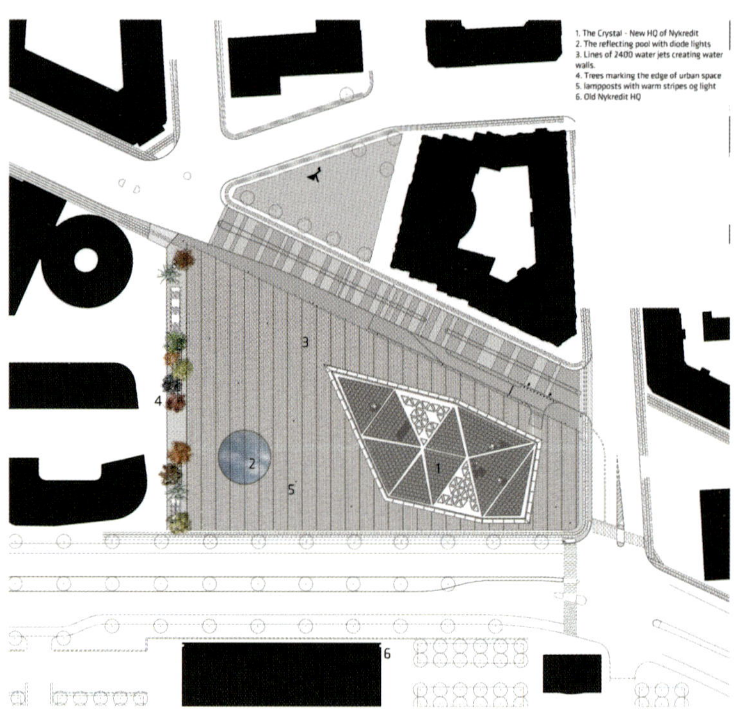

项目地点：
丹麦哥本哈根
面积：
5500 平方米
团队：
Stig L. Andersson, Hanne Bruun M ller, Signe Hertzum, Malin Blomquist
合作者：
Schmidt Hammer Lassen
设计公司：
SLA

Copenhagen lies by the water. In recent years it has become more attractive to live and work by the water than it was, when the harbor was the city's backside with industry and heavy transportation. Many new and local neighborhoods have emerged after the industry and storage areas have been dismantled and removed. But water is also a growing problem in Copenhagen: rain water. Climate changes have increased pressure on the inadequate sewerage in the city. When it rains, the sewers are threatened by overload and the polluted sewage water ends up in the harbor. Water is both an amenity to the city and a testimony to how important it is that we take care of the water's qualities through an understanding of its harmful as well as its life-giving and sensual properties.

哥本哈根是依存于水资源利用与开发的城市。以前由于重型工业运输污染，海港地区成为城市生活环境的落后区，而近些年来，这里相比以前吸引了越来越多的人们过来工作和生活。很多新来的和本地的居民在工业区、仓储区拆除移走之后迁入到这里。但是，城市污水治理仍然是哥本哈根日益严重的问题：譬如，雨水。气候变化给城市本不完善的污水再生利用的处理设施带来了越来越多的压力。当下雨的时候，污水排水道会有过载的危险，而且受到污染的污水也将排放到海港中。可见，水既是城市形象的美容师又是市民生活健康的医疗师，通过了解水的危害性和水赋予生命的灵性财富，我们懂得了水质对于人们生活发展的重要性。

The urban space CLOUD lies on the border between the old center of Copenhagen and the modernist harbor, where the old city's apartment buildings, with faceted and rough exterior surfaces, meet the smooth and reflective business domiciles in a new spatial order: An order where the classic axial and hierarchical room-partition turns into the liquid, non-hierarchic spatial sequence.

CLOUD 城市空间位于哥本哈根的老中心区和现代海港之间的交界处，其中老城区的公寓建筑物建有雕琢切割过的外表面，呼应了商业住宅追求平滑反光效果的新式空间原则：这项原则是基于将有古典轴线、分层次的室内隔断转变成流线型、不分层的空间序列。

It is from these traits, the urban space CLOUD originates. Here, The Crystal – the new HQ of a major Danish bank – rises like frozen water and reflects the sunlight with its smooth, sharp and jagged surface. On days with overcast weather, the building draws the gray shades in and combines with the ambient colors and expressions of the weather.

CLOUD 城市空间的开创正是基于这些特点。在这里，水晶大厦——一个丹麦重要银行的新总部——外表面好似结冻的水面，能够利用其光滑、锋利、凹凸不平的表面反射阳光。到了阴雨天的时候，大厦仿佛笼罩了灰色的阴影，并且融合了周围环境的色彩和天气的变化。

All in all, CLOUD provides Copenhagen with a sensuous and intimate urban space that not only provides local acclimatization and amenity values, but also allows the residents of Copenhagen to enjoy some of their city' s most characteristic traits: clouds, rain and mist.

总而言之，**COULD** 设计为哥本哈根提供了感性而亲切的城市空间，这里不仅能够提供当地健康舒适的环境，而且能够让哥本哈根的市民享受到自己城市的独具特色的一面：云、雨和雾。

中邦城市雕塑花园
Zobon City Sculpture Garden

LOCATION:
Shanghai, China
TEAM:
Kui-Chi Ma, Ying-Yu Hung, Gerdo Aquino, Kui-Chi Ma, Haven Kiers, Conard Lindgren
CLIENT:
Edward Jiang, CEO, Landmark Development Consulting (Shanghai) Co., Ltd.
ARCHITECTS:
Arquitectonica Architects

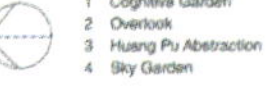

项目地点：
中国 上海
团队：
Kui-Chi Ma, Ying-Yu Hung, Gerdo Aquino, Kui-Chi Ma, Haven Kiers, Conard Lindgren
委托人：
Edward Jiang, CEO, Landmark Development Consulting (Shanghai) Co., Ltd.
建筑师：
Arquitectonica Architects

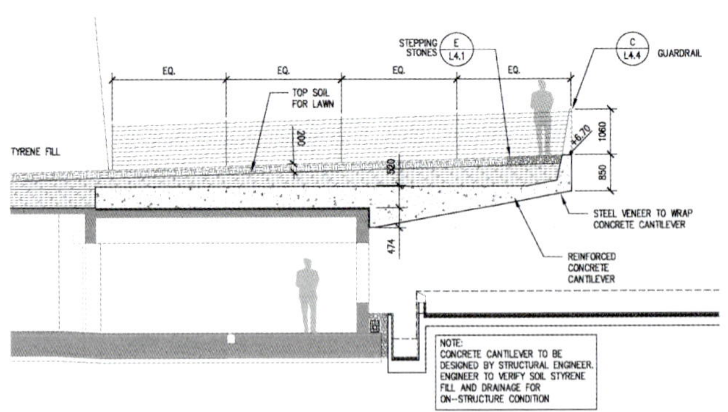

The Zobon City Sculpture Garden lies at the center of a 5,000 unit multi-family residential infill development in the Pudong district of Shanghai, China. The design objective was to create an innovative model for multi-family habitation that integrates art, landscape and architecture in ways that make dense, urban living more sustainable. On a mere 0.6 hectare site, the landscape architecture expresses three moments which celebrate the inherent, and often invisible, beauty of the city.

中邦城市雕塑花园坐落于上海浦东新区一处有5000个单元多户填入式开发住宅区的中心。设计目标是为多户式住宅提供一个创新的典范，运用使密集的城市居住环境更加可持续的方式将艺术、景观和建筑有机地结合起来。花园仅占地0.6公顷，从三个侧面烘托出城市内在而经常察觉不到的美。

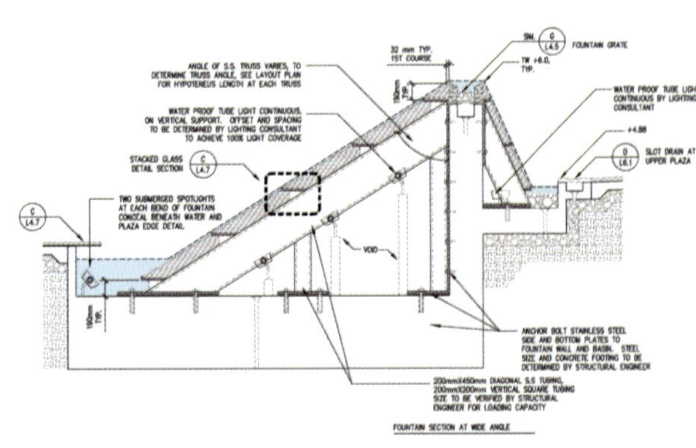

Project Narrative

As an avid art collector, the client was interested in a landscape/open space program that would provide a creative outlet for urban residents and visitors. The notion of landscape as art became the driving force behind the design of the Zobon City Sculpture Garden. The basis for landscape "art" was derived from the landscape architect's understanding of the city and the way in which it interacts with natural forces. Urban phenomena such as flooding, land subsidence and freeway infrastructure are all central to the of environment Pudong Shanghai. The landscape design attempts to capture and reinterpret these urban/natural systems dynamic by creating three distinct moments within the project.

项目介绍

作为一名十分投入的艺术品收藏者，对通过景观及开放空间设计为居民和游客提供一个有创造力的平台有着很大的兴趣。把景观当作艺术的理念是中邦城市雕塑花园景观设计的推动力。景观艺术的根基来自于景观设计师对于城市和它与自然力量之间相互关系的理解。城市中的一些现象，如洪水、地面下沉和高速公路的基建对上海浦东环境发展的关键问题。景观设计试图通过打造三个侧面抓住并重新解释这些城市及自然系统的本质情况。

商业广场景观
BUSINESS SQUARE LANDSCAPE

ChonGae 运河广场

ChonGae Canal Restoration Project

LOCATION:
Seoul, Korea
PHOTOGRAPHER:
Taeoh Kim
DRAWINGS/RENDERINGS:
Mikyoung Kim Design
DIMENSIONS:
91,000 m^2

项目地点：
韩国汉城
摄影师：
Taeoh Kim
绘图 / 效果图：
Mikyoung Kim Design
尺寸：
91 000 平方米

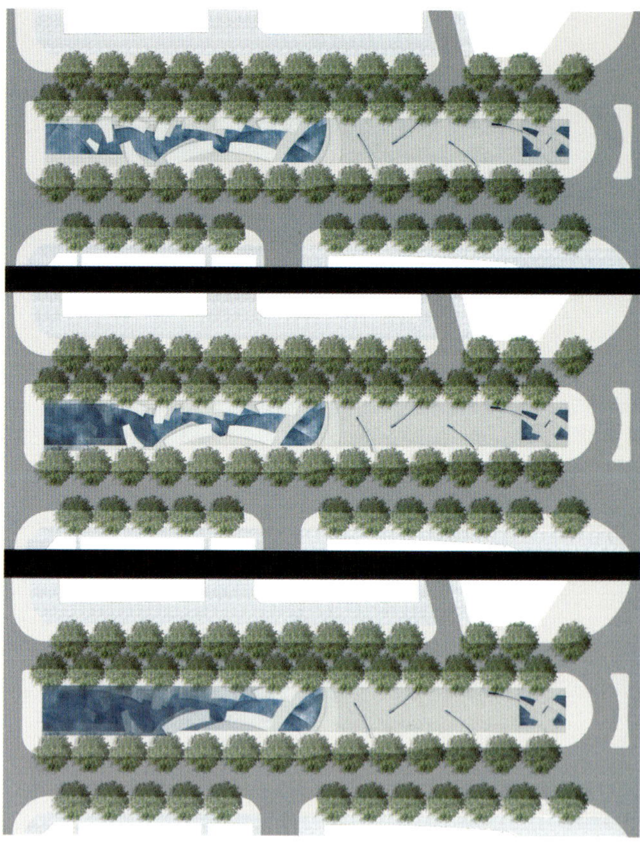

The ChonGae Canal Restoration Project is an ambitious redevelopment initiative that transformed the urban fabric of Seoul, Korea. This design was the winning project in an international competition and celebrates the source point of cleansed surficial and sub grade runoff from the city at the start of this seven mile green corridor. The main competition requirement was to highlight the future reunification of Korean Peninsula. The project symbolizes this political effort through the use of donated local stone from each of the eight provinces of Korean Peninsula. The individual stones act to frame the urban plaza and the eight source points where runoff is daylighted and represents the unified effort in the transformation of this urban center.

ChonGae 运河修复工程是一项规模宏大的改造计划，改变了韩国首尔的城市结构。该设计是一项国际竞赛的获奖作品，有力地表达出这个七英里的绿色长廊的源头表层和亚表层清澈的径流。竞赛主要的要求是强调出朝鲜半岛未来的统一。这项工程具有政治意义，使用了从朝鲜半岛八个省捐来的石材。这些石头一块一块构筑起了广场和八个径流源点，这几个源点暴露在自然光下，在这个城市中心的改造中象征着统一。

Project Narrative:
The ChonGae River Restoration Project is located at the important source point of this seven-mile green corridor that begins in the central business and commercial district of the city. The goal was to restore this highly polluted and covered water-way with the demolition of nearly four miles of at grade and elevated highway infrastructure that divided the city. The outcome is the creation of a pedestrian focused zone from this former vehicular access way that brings people to the historic ChonGae River while mitigating flooding and improving water quality.

ChonGae 运河修复工程位于这个七英里绿色长廊的源头处，就在城市商业中心。目标是拆除近四英里长、把城市分开的水平突起的公路基建，恢复这个高度污染和被覆盖的水道。成果就是把先前这个车道改造成了步行区，吸引人们来到具有历史意义的 ChonGae 河，同时减轻了水灾风险，改善了水质。

The charge of this international design competition was to create a symbolic representation of the future reunification of Korean Peninsula within a highly active public plaza. This winning proposal defined the eight provinces through the use of local materials and eight sources of water. Regional stone quarried from each of the eight areas, eight source points of water and fiber-optic light highlight this collaborative effort of reunification and restoration.

这个国际设计竞赛的要求是通过一个非凡的公众活动广场来象征朝鲜半岛未来的统一。获奖的提案通过利用当地的材料和八个水源点来体现八个省区。从八个地方采来的石材、八个水源头和光纤照明突出体现了这个象征统一、多方参与的修复工程。

The design was guided by the water levels from hour to hour and season to season, while addressing the catastrophic flooding that occurs during intense storms in the Monsoon season. The unique sloped and stepped stone elements allow for a reading of the various levels of water while encouraging direct public engagement with the river. The restoration of this area is the first step in a major redevelopment effort of this seven-mile river and current ambitious architectural redevelopment projects that frame this natural drainage basin to the city. Since the ribbon cutting ceremony in October 2005 on the main plaza, nearly 10 million visitors and residents have visited the river.

设计考虑了不同季节、不同时刻的水位变化，同时应对了雨季时暴雨引发洪灾的问题。石材铺建的带有独特的坡度和阶梯，是考虑到了水位的不同变化，还吸引人们直接去接触水。这个区域的修复是整个项目的一期工程，黄土里长河的重建也是当前这个水流区域多个改造项目中的一个。从 2005 年 10 月在主广场剪彩到现在，已经有近一千万人次的居民和游客来参观。

대~한민국
DONGWHA
한국 축구
HEAD

In addition to the environmental restoration effort, this urban open space has become a central gathering place for the city which is in dire need of more public landscapes. The Class II water quality level has allowed for families to come and reengage with this historical river. During specialized events such as the traditional New Year's festivals, political rallies, fashion shows and rock concerts both the plaza and the Water Source area get redefined in inventive ways. Recently, coins that were tossed in the canal by visitors were collected from the basin and thousands of dollars were donated to local charities.

除了对环境的恢复外，这里也成了城市急需的一个重要公共景观聚集地。水质达到了2级标准，就是为了能让全家老小来这条历史河流边放心地游玩。广场和径流源头区在特别的时候能发挥出不同的功能，比如传统的新年节日、政治集会、时装表演、摇滚音乐会。最近，将参观者投在河里的硬币从河底收集起来，足有数千美元之多，这些钱被捐给了当地的慈善组织。

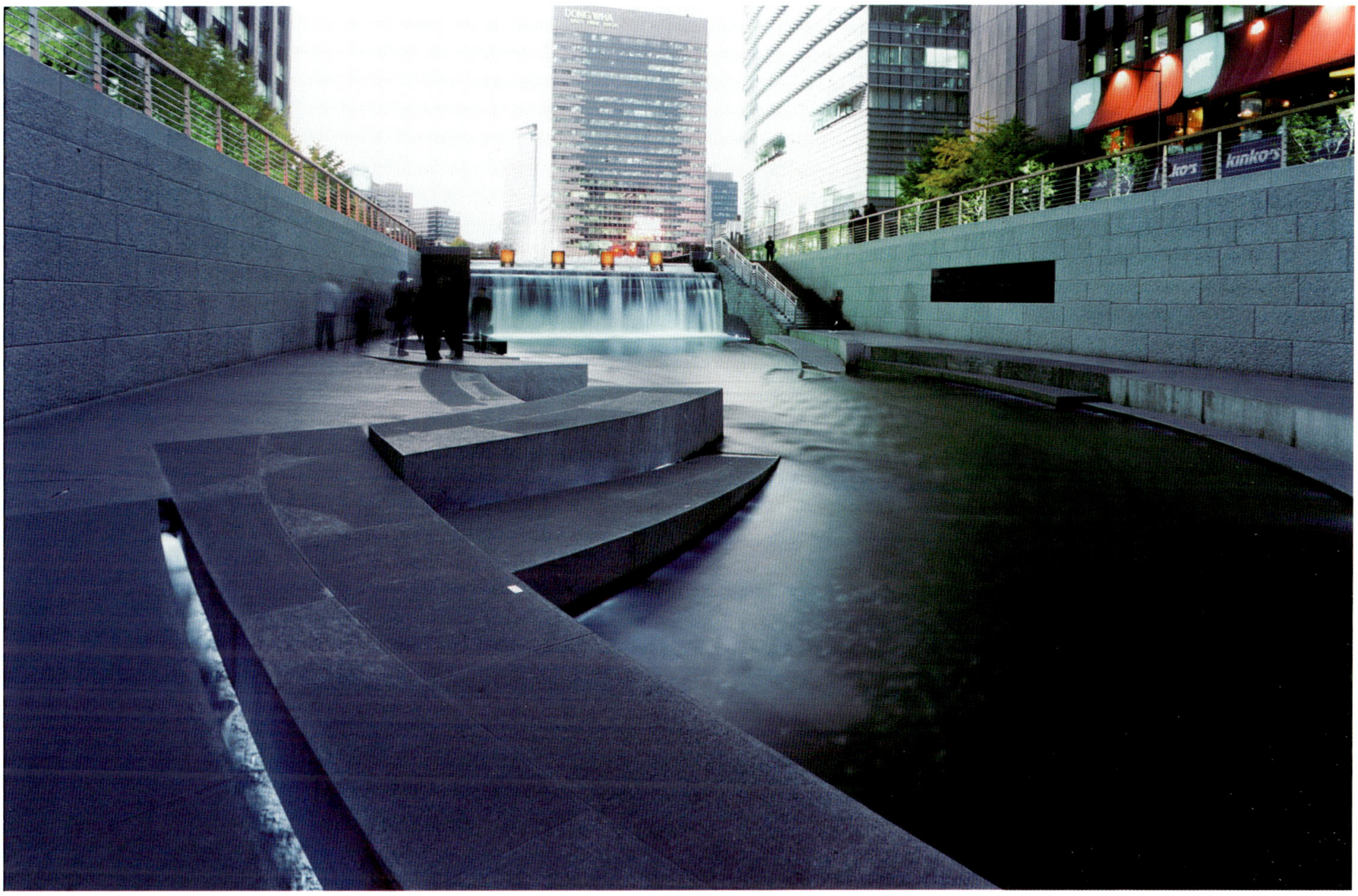
DONG WHA
kinko's

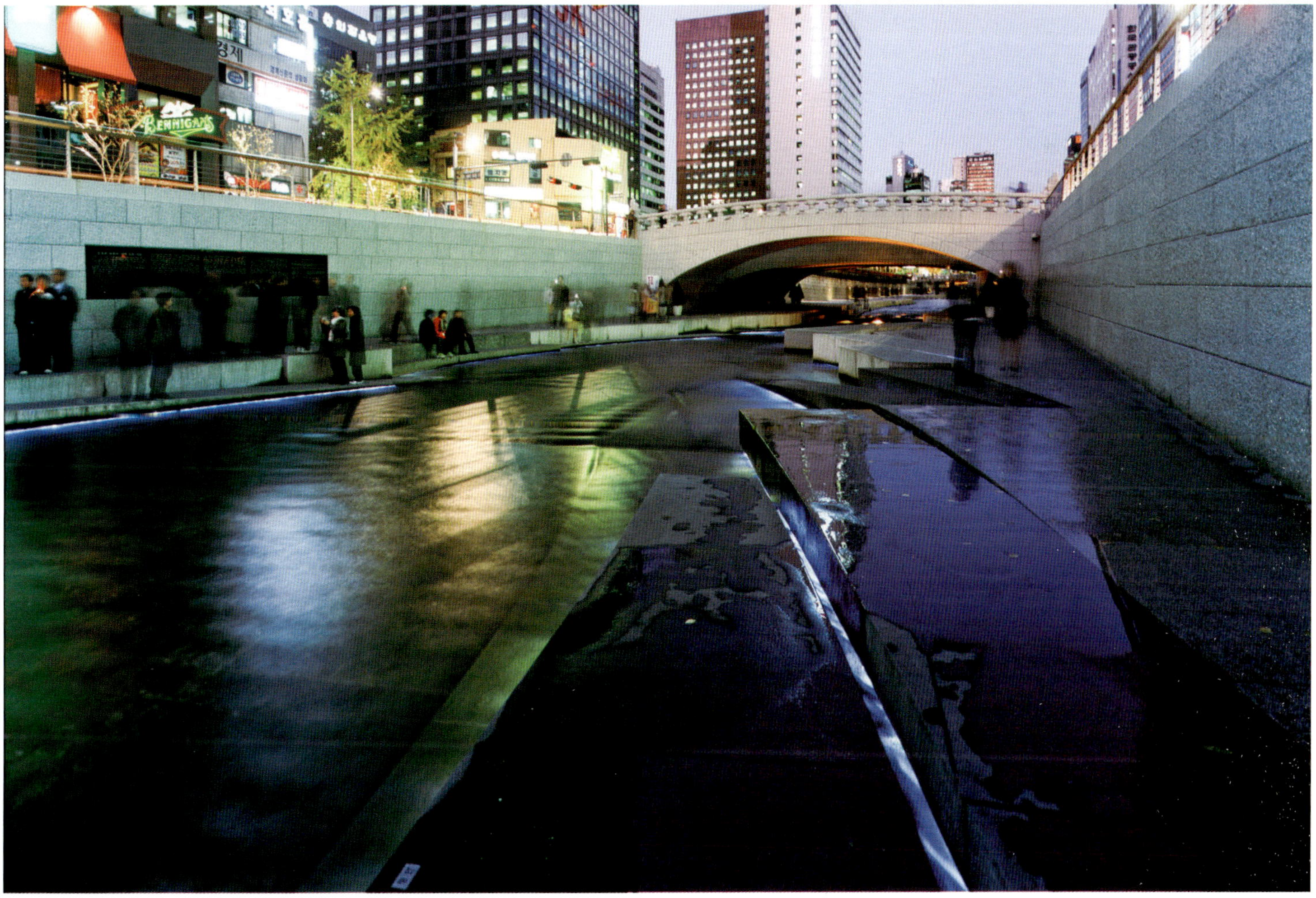
BENNIGAN'S

附子广场
FUZI Plaza

LOCATION:
Austria
LANDSCAPE ARCHITECT:
Otto Kapfinger

项目地点:
澳大利亚
景观设计师:
Otto Kapfinger

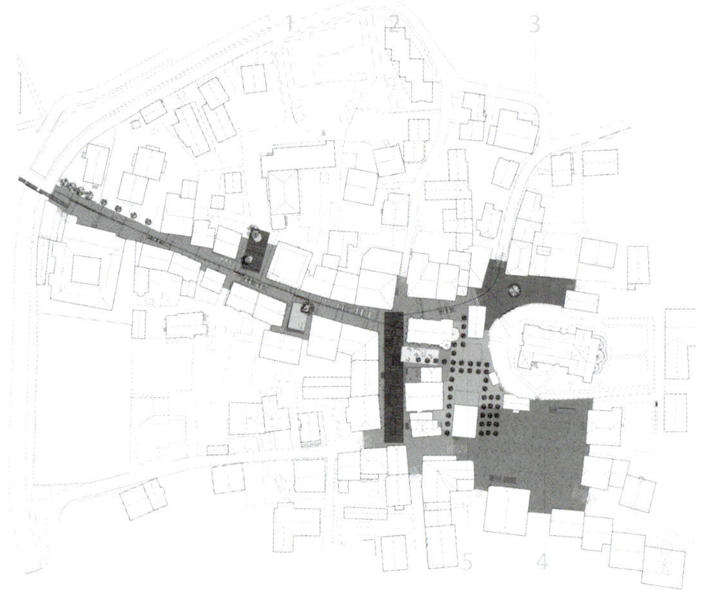

Conversions of old city centers into traffic-free shopping and tourists' malls have frequently been performed according to trite recipes. The shops' increase in turnover mostly stands in opposite to the commercial alienation of old structures – a process of "gentrification" in which old substance is scenically pepped up, while local societies are hollowed out and superseded. Contrary to the interpretation of the task, the commission itself developing from a small local competition was not at all typical of AllesWirdGut, declared freaks for cars as they are. Innichen experiences tourist peaks. In the off-season, the center is deserted, and the locals suffer from a hangover following the seasonal stress. The center's novel open space development reacts to these seasonal fluctuations. The varied zones are interactive and can easily be modified. Wooden grates set upon new event platforms are removed after the season proper and replaced by flowerbed trays. Alternately, individual panels can also be flooded with water, and the vacant spaces that are "too large" in the off-season are given useful arrangements complemented by a wellness factor. Moreover, resourceful technologies allow the various new surfaces to be undertaken very inexpensively, thus releasing budgets for the "alternating phases".

老旧城市中心向无车购物和旅游中心改造一般都会遵循传统的套路。店铺营业额的增加通常都与老旧建筑商业化差的情况相对立——这是"中产阶级造成的恶果"：老旧的物品被注入活力，当地的社会被掏空、替代。与对本次任务的解读相反，从一次小规模当地设计竞赛中拿到的项目本身就不符合对汽车有着狂热爱好的 AllesWirdGut 的原有作风。Innichen 经历了旅游高峰。在旅游淡季，旅游中心就被荒废了，当地人就要经历这种高峰后的松垮感，紧接着就是季节性的压力。这个购物旅游中心新颖的开阔空间的设计与这种季节性的变化相互呼应。多样化的区域相互作用，可以轻易地进行更改。活动平台上的木条可以在高峰期过后拆掉，换成花盆。另一方面，单个嵌板里也可以注满水。对于在淡季显得"过大"的未占用的空间也进行了妥当的安排，并加上了一个保健元素。不仅如此，利用巧妙的技术可以在节约成本的情况下实现这些更改，为"交替阶段"节省了预算。

Bank
für Trient und Bozen
GRAUER BÄR
ORSO GRIGIO
HOTEL
RESTAURANT
HOTEL
RESTAURANT
BAR
BAR

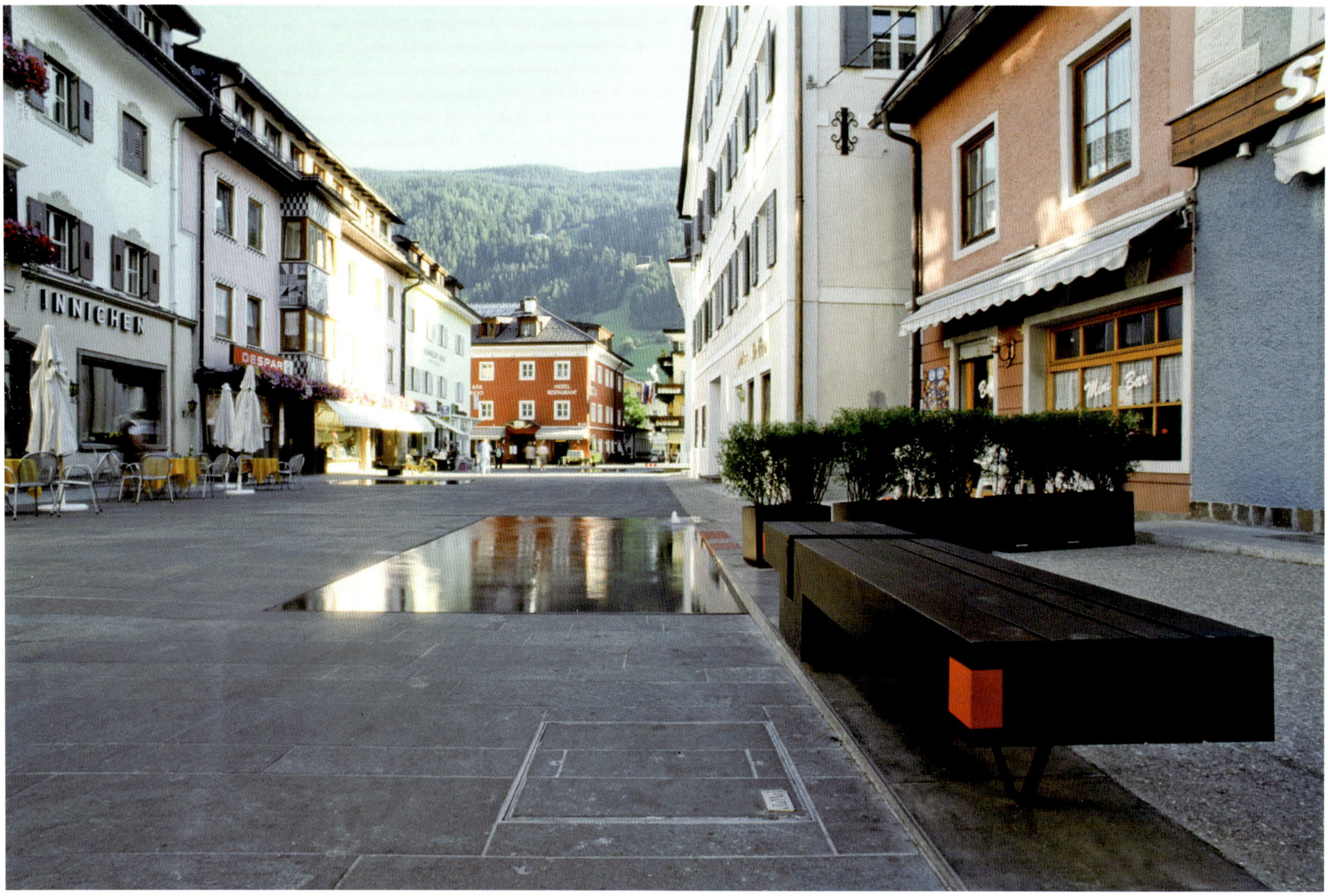
INNICHEN
DESPAR
Mini Bar

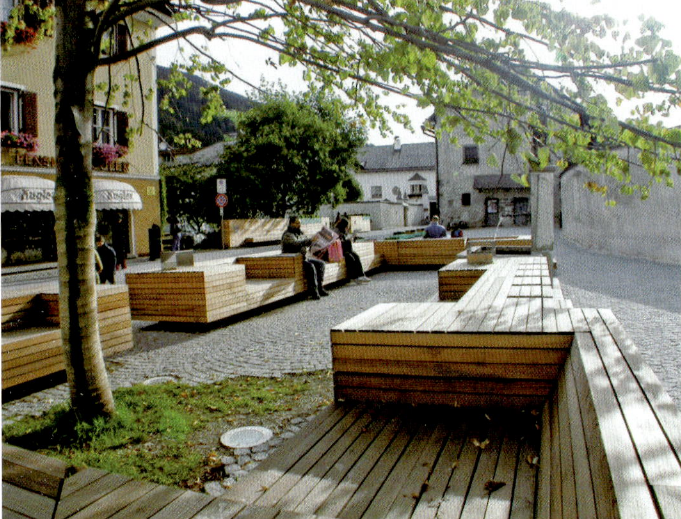

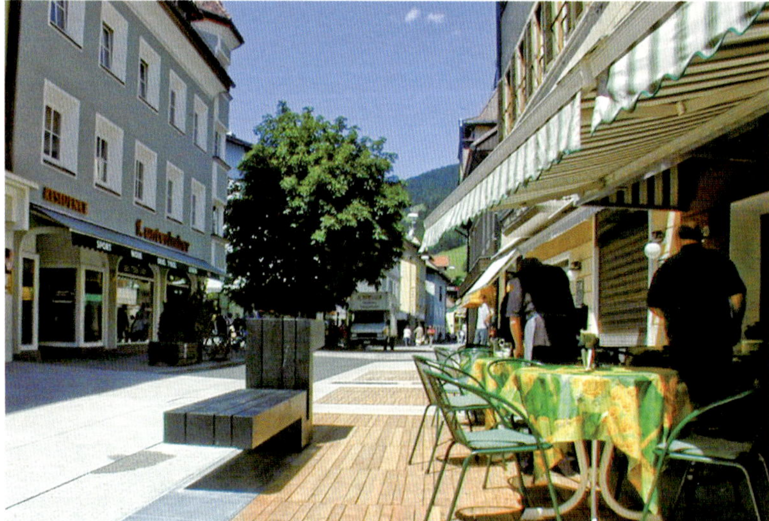

GRAUER BÄR
ORSO GRIGIO

建筑节“零浪费馆”广场

Archifest Zero Waste Pavilion

LANDSCAPE ARCHITECT:
WOW Architects
WORK TEAM:
Fernando Velho, Prabhu Sugumar, Christopher Lee
Liang Neng, Yvonne Yung
LOCATION:
Singapore
AREA:
$270m^2$(Site), $40.7m^2$(Length of Pavilion)
PHOTOGRAPHER:
Aaron Pocock, C3M Studio

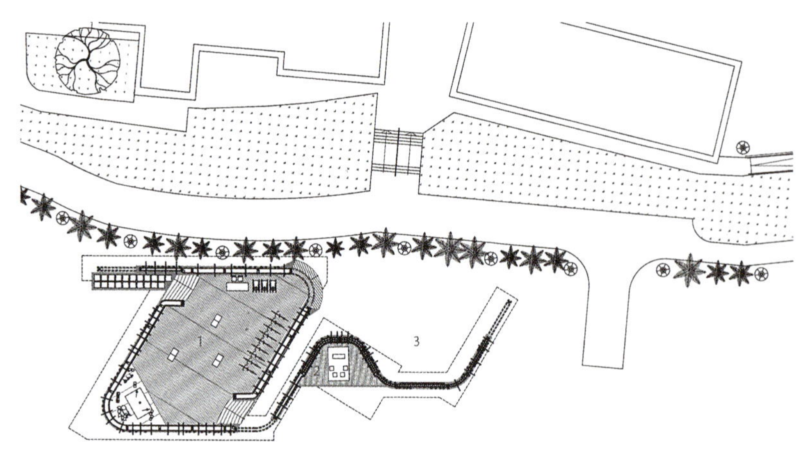

景观设计师：
WOW 建筑事务所
工作团队：
Fernando Velho, Prabhu Sugumar, Christopher Lee、梁能、翁虹
项目地点：
新加坡
面积：
$270m^2$（场地）；$40.7m^2$（展馆长度）
摄影师：
Aaron Pocock、C3M 工作室

As the winning entry for the Archifest 2012 Pavilion Competition, the Zerowaste Pavilion was a collaborative effort by WOW Architects and the Archifest organizers to create a flexible event space that also embodied the festival's central theme of "Rethinking Singapore". By reusing materials in a new way to extraordinary functions and delight, the Pavilion manages to engage and inspire participants and visitors in line with the thematic program of Archifest.

“零浪费馆”是 2012 年建筑节展馆竞赛的获奖者，它是由 WOW 建筑事务所和建筑主办单位合作完成的。他们打造了一个灵活的活动空间，象征着艺术节的中心主题——“反思新加坡”。展馆通过对材料进行再利用，号召参与者和游客遵循建筑节的主题项目，来获取独特的功能和乐趣。

General view

Central to our approach for this pavilion was to create a space that had maximum impact on the Archifest participants, but minimal impact on the site and overall environment. This goal to utilize "Zero-waste" construction methods became our guiding principal throughout the design process, detailing, fabrication, on-site installation, and lastly, the dismantling and "Afterlife" of the materials used to construct the pavilion.

我们的主要目的是创建一个对艺术节参与者有着重要影响的空间，同时将其对场地和整体环境的影响降到最低。这个采用"零浪费"施工方法的目标成了我们在设计、制图、制作、现场安装以及最后的拆卸过程中的指导原则，在处理建馆材料的“后世”时，我们也同样遵循着这个原则。

韦斯利区道路和公共空间广场

Wesley Quarter — Laneway and Public Realm

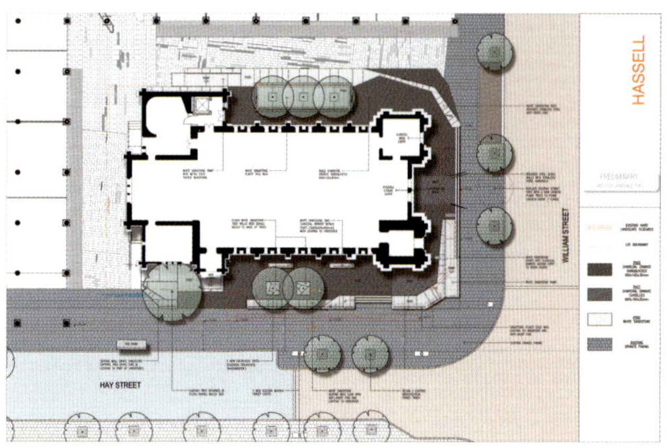

LOCATION:
Australia
CLIENT:
Uniting Church of Australia

项目地点：
澳大利亚
委托人：
Uniting Church of Australia

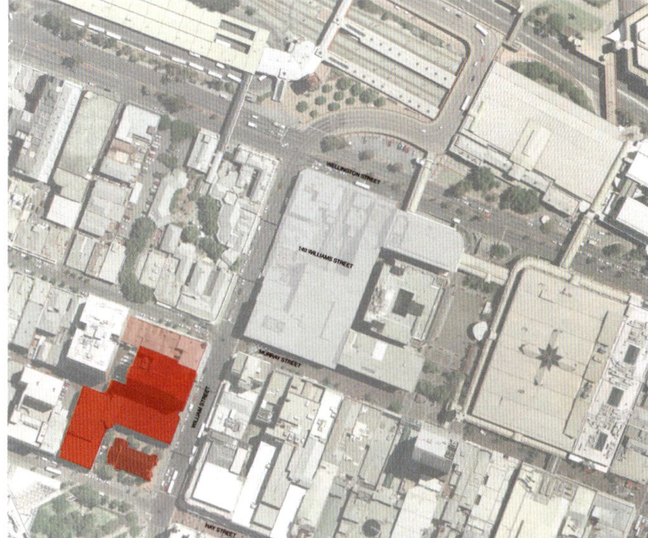

The upgrade to the public spaces surrounding the historic Wesley Church was a key component of a collaborative approach to transform the heritage Wesley Buildings into a vibrant commercial and retail center within the heart of the city. The spaces are dynamic in design and materiality. They focus on pedestrian comfort and assist in elevating Wesley Quarter as a cultural and fashion icon of Perth.

在将韦斯利大楼改造成零售商业中心的一系列方案中，对历史著名的韦斯利教堂周边公共区域的改造是重要的一部分。空间的设计和所采用的材料都很有动感。设计注重行人的舒适性，为将韦斯利区提升为的佩斯文化和时尚标志做出了贡献。

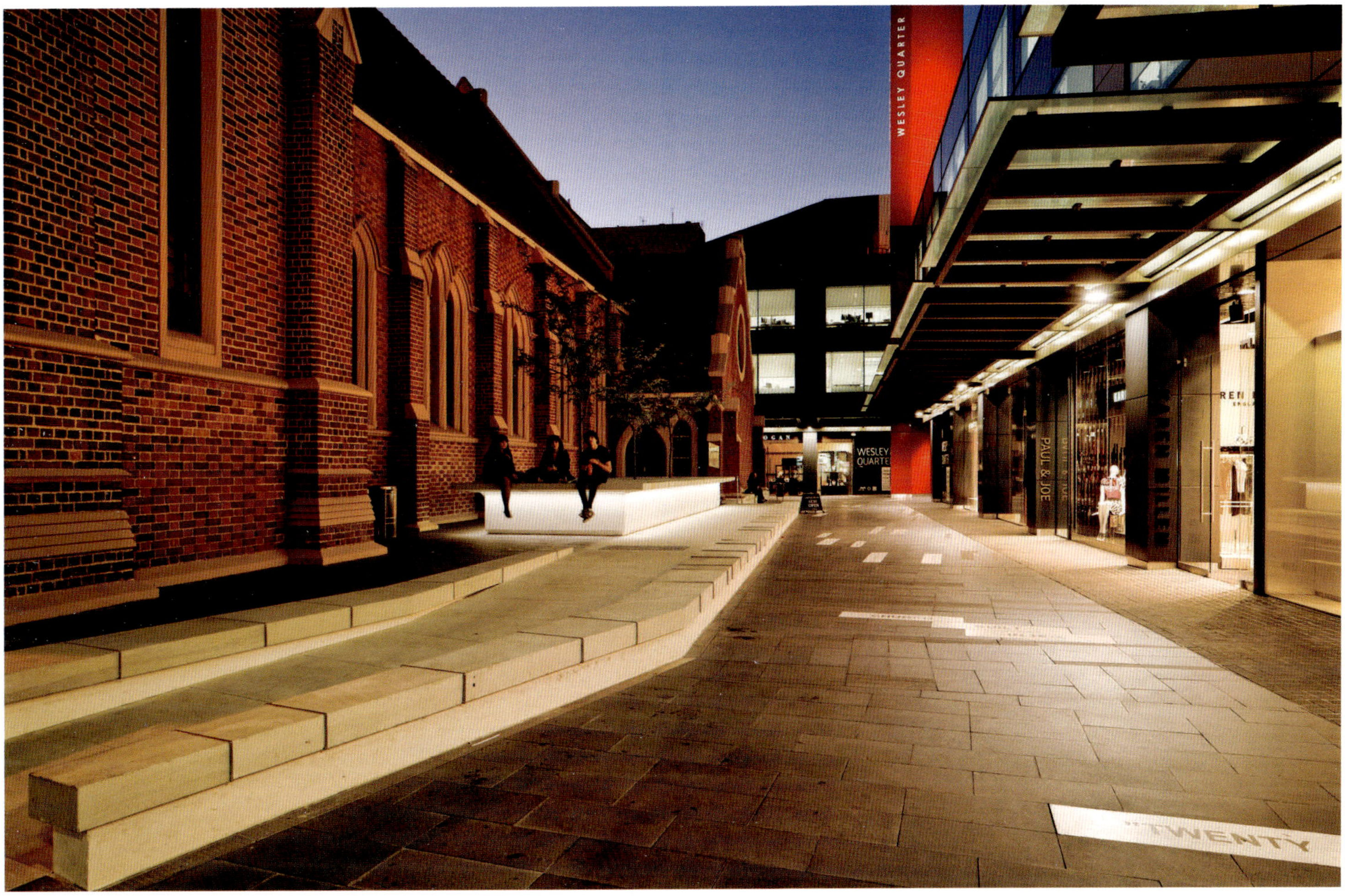
WESLEY QUARTER
PAUL & JOE
TWENTY

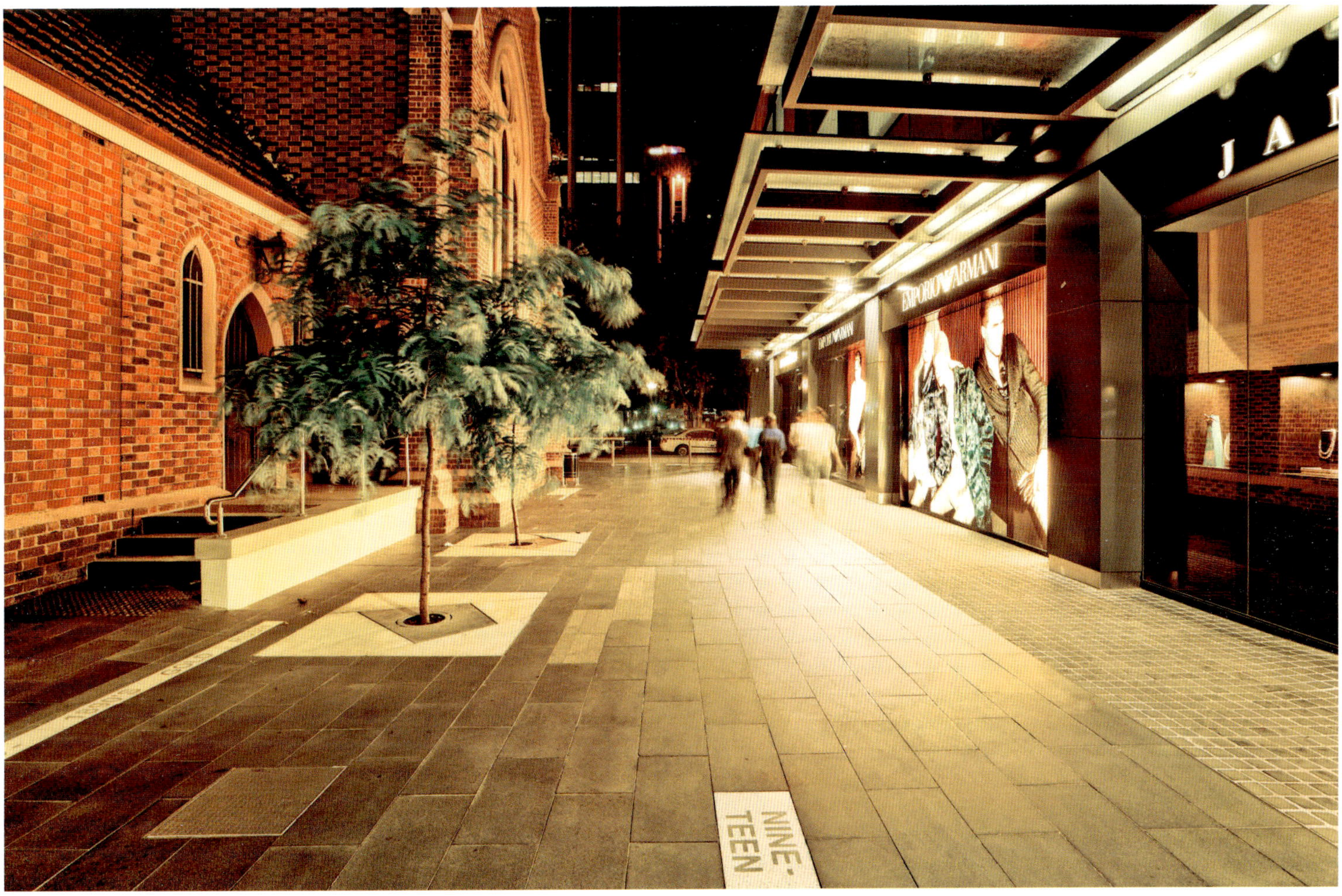
EMPORIO ARMANI
NINE-
TEEN

The Wesley Quarter is a sophisticated urban response to city using quality materials in a highly constrained and culturally significant heritage site. It contributes to the image of the city by respecting its historical and cultural context whilst providing a contemporary counterbalance to the grittiness of public life.

在高度内敛并具有文化重要性的遗址内，韦斯利区是对采用高品质材料的城市的一个高端的回应，通过对其历史和文化背景的尊重，对提升城市的印象起到了重要作用，同时也为枯燥的公共生活注入了现代元素。

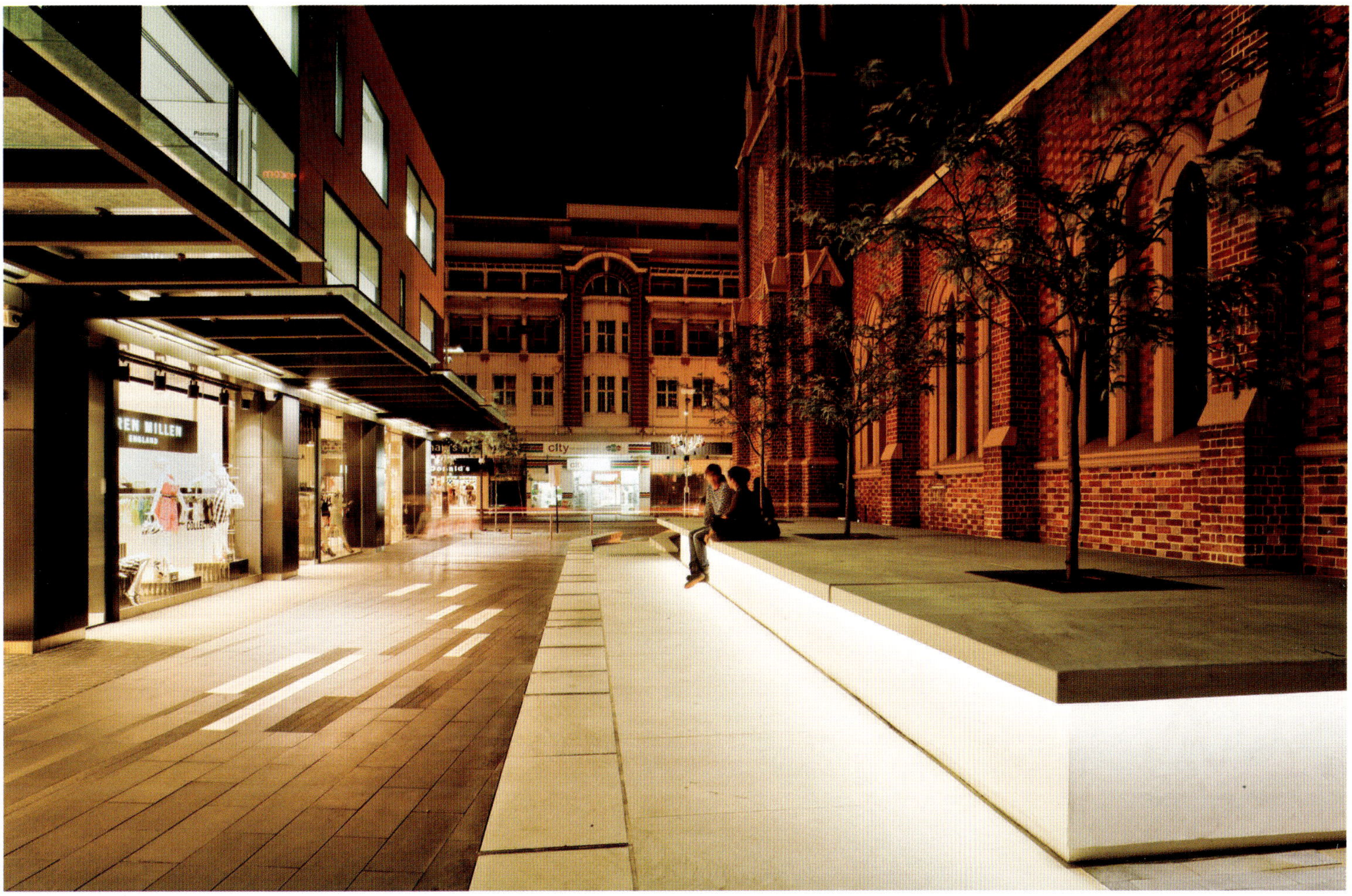
REN MILLEN
ENGLAND
city

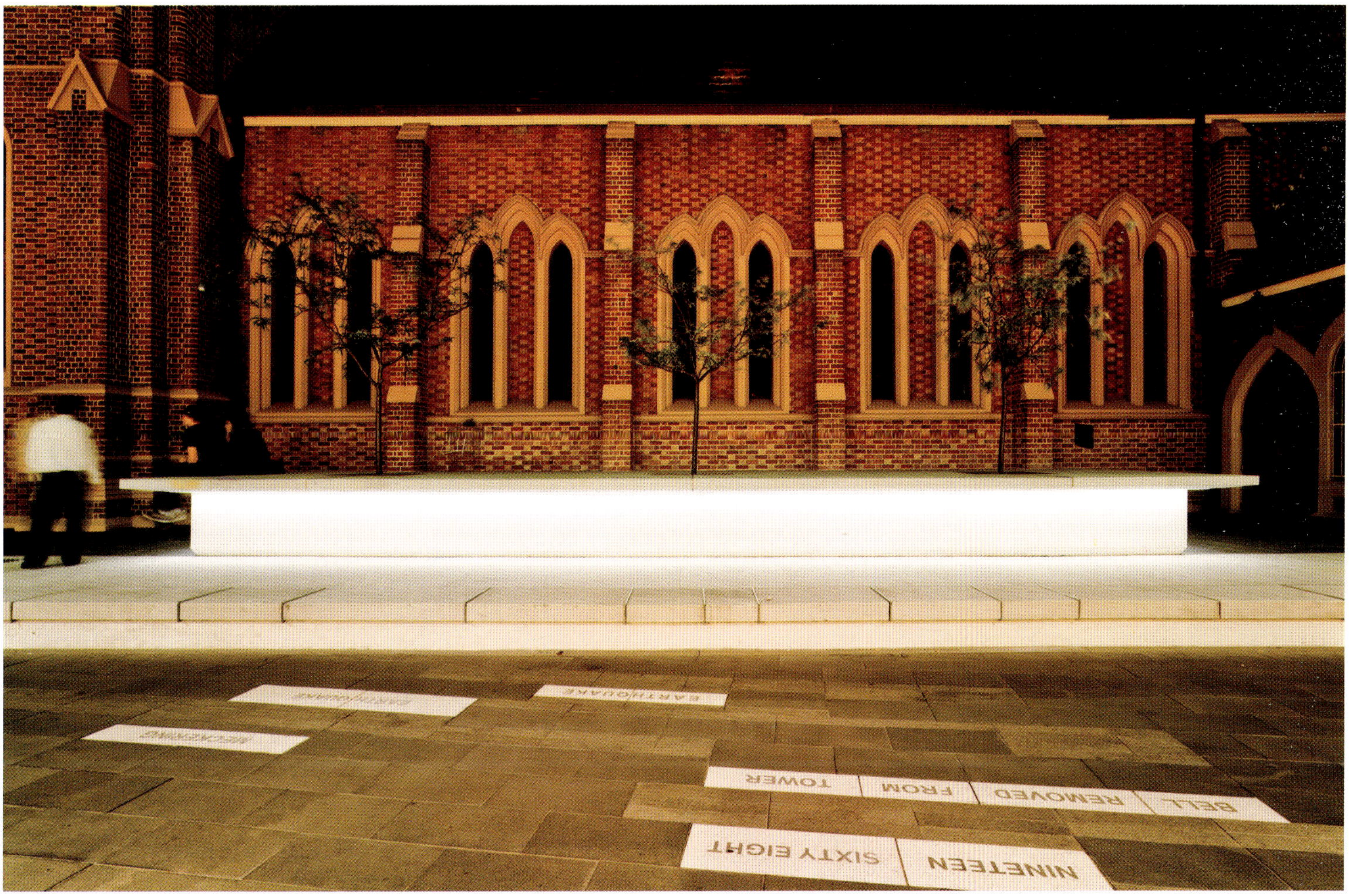
EARTHQUAKE
BELL REMOVED FROM TOWER
NINETEEN SIXTY EIGHT

城市花园广场
City Square Park

LOCATION:
Singapore
PHOTOGRAPHER:
See Chee Keong
AWARD:
BCA Universal Design Award for Built Environment, Gold

项目地点：
新加坡
摄影师：
See Chee Keong
奖项：
BCA 全球环境建设设计奖，金奖

City Square Urban Park fronts City Square Mall's entrance and provides a recreational environment for nearby residents and visitors alike. In addition, the natural space serves as a constant reminder of the importance of environmental conservation.

城市广场都市花园面朝着城市广场购物中心入口处，为附近的居民和游客提供了一个休闲的环境。而且，这个自然空间还起到了不断提醒环境保护重要性的作用。

The sunken plaza is the focal point of the park, serving as a multi-purpose space while also acting as a passageway linking the MRT station to the mall. An eco-roof – comprising of solar panels, low-E glass panels and a green roof – traps naturally produced energy that in turn powers the lightings, regulates temperature and controls wind circulation.

这个凹陷的广场是公园的焦点，是个多功能空间，而且还是连接 MRT 车站和购物中心的通道。一个由太阳能板、低辐射玻璃板和一个绿色顶盖构成的生态屋顶能采集自然界的能量，然后反过来为照明提供动力，调节温度和控制风的循环。

Eco-friendly materials, such as Eco-tiles and recycled timber, were used to construct the park' s various structures, such as the playground.

环保材料，比如生态瓷砖和再生木材，被用来建造花园的不同结构，比如运动场地。

Balgowlah 村庄广场

The Village, Balgowlah

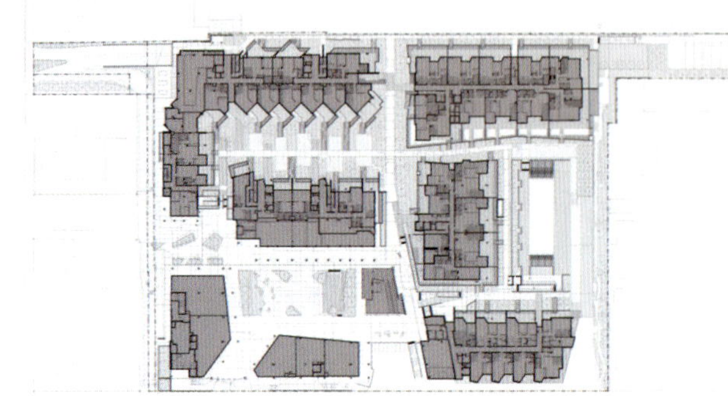

CONSTRUCTION SITE:
197-215 Condamine Street
Balgowlah
NSW 2093
Australia

BUILDING AREA
45,715m^2

LAND AREA
21,000m^2

MAIN DESIGNERS
Reg Smith – Project Principal
Keith Cottier – Urban Design
Steve Black – Project Director
Michael Buchtmann – Residential Design
Rob Doak – Retail Design
Jane Johnson – Architect
Pip Bowling – Architect

工程场址：
澳大利亚新南威尔士州（2093）Balgowlah 区 Condamine 大街 197–215 号

建筑面积：
45 715 平方米

占地面积：
21 000 平方米

主设计师：
Reg Smith– 工程负责人
Keith Cottier– 城市设计
Steve Black– 工程主管
Michael Buchtmann– 住宅设计
Rob Doak– 店铺设计
Jane Johnson– 建筑师
Pip Bowling– 建筑师

The Village is a new mixed-use development located on a 2.1-hectare site in the northern Sydney suburb of Balgowlah. Situated on an elevated position, it has a view out to the northern Sydney coastline, a hub of recreational activity for many local residents and visiting tourists. The Village also boasts proximity to the beaches and was designed to reflect the casual and relaxed lifestyle that living in the area affords. The site includes the original 1960s Totem Shopping Centre, which has been redeveloped and integrated as part of a larger retail project, Stockland Balgowlah. A key aspect of the shopping centre' s transformation was to evoke the atmosphere of a town square to encourage public accessibility and mobility.

这个村庄是一个新的混合用途开发项目，位于悉尼北部郊区 Balgowlah，占地 2.1 公顷。因为场址地势高，从这里可以眺望悉尼北部海岸，那里是当地居民和游客的一个休闲中心。村庄引以为豪的是离海滩近，它的设计反映出了当地那种休闲放松的生活方式。场址包括建于 20 世纪 60 年代的图腾购物中心，该中心几经重新开发改造，是 Stockland Balgowlah 一个更大的购物工程的组成部分。购物中心整改的一个关键点是营造城市广场的氛围，方便公众使用和流动。

Stockland
BERKELOUW
BOOKS

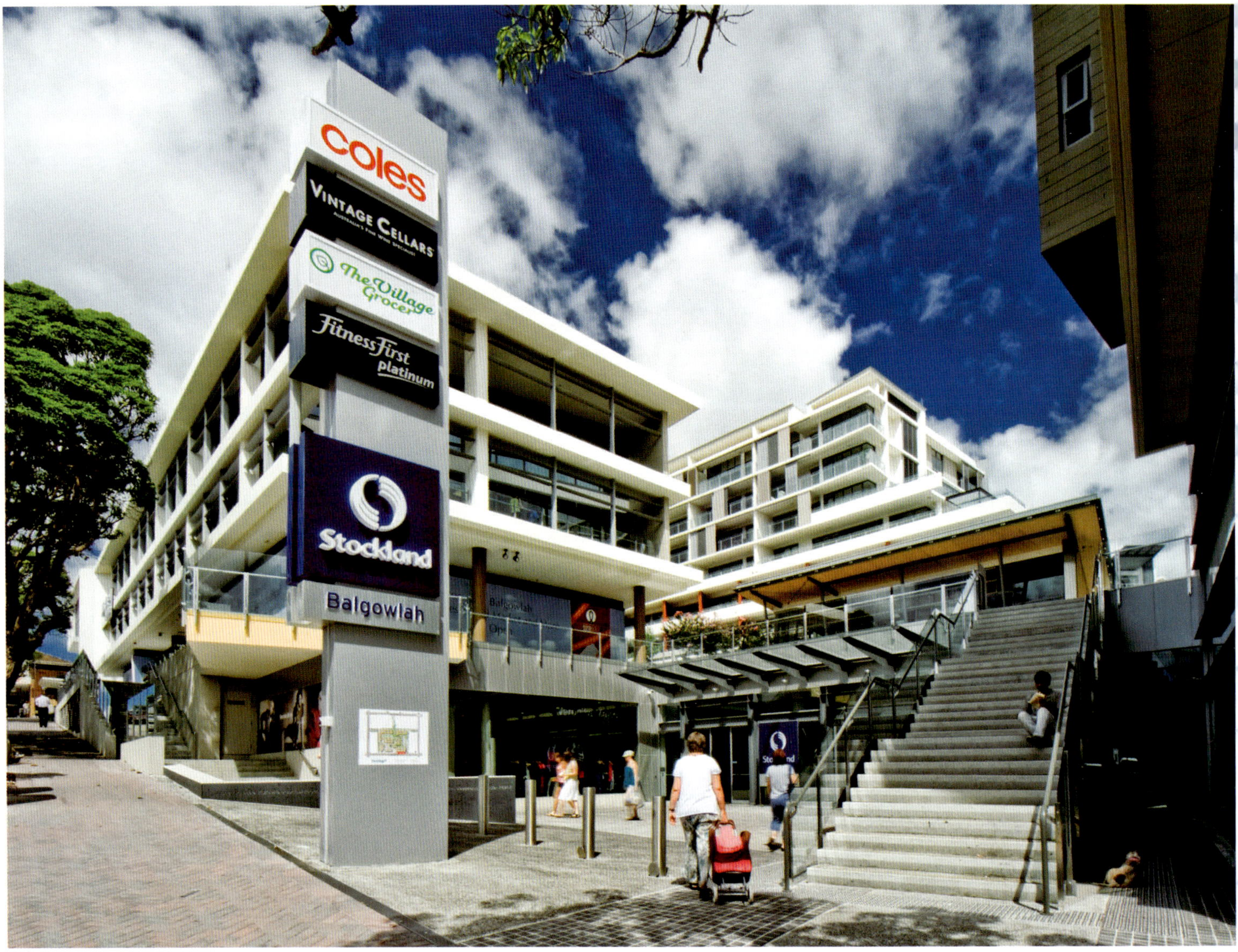
coles
VINTAGE CELLARS
The Village Grocer
Fitness First
platinum
Stockland
Balgowlah

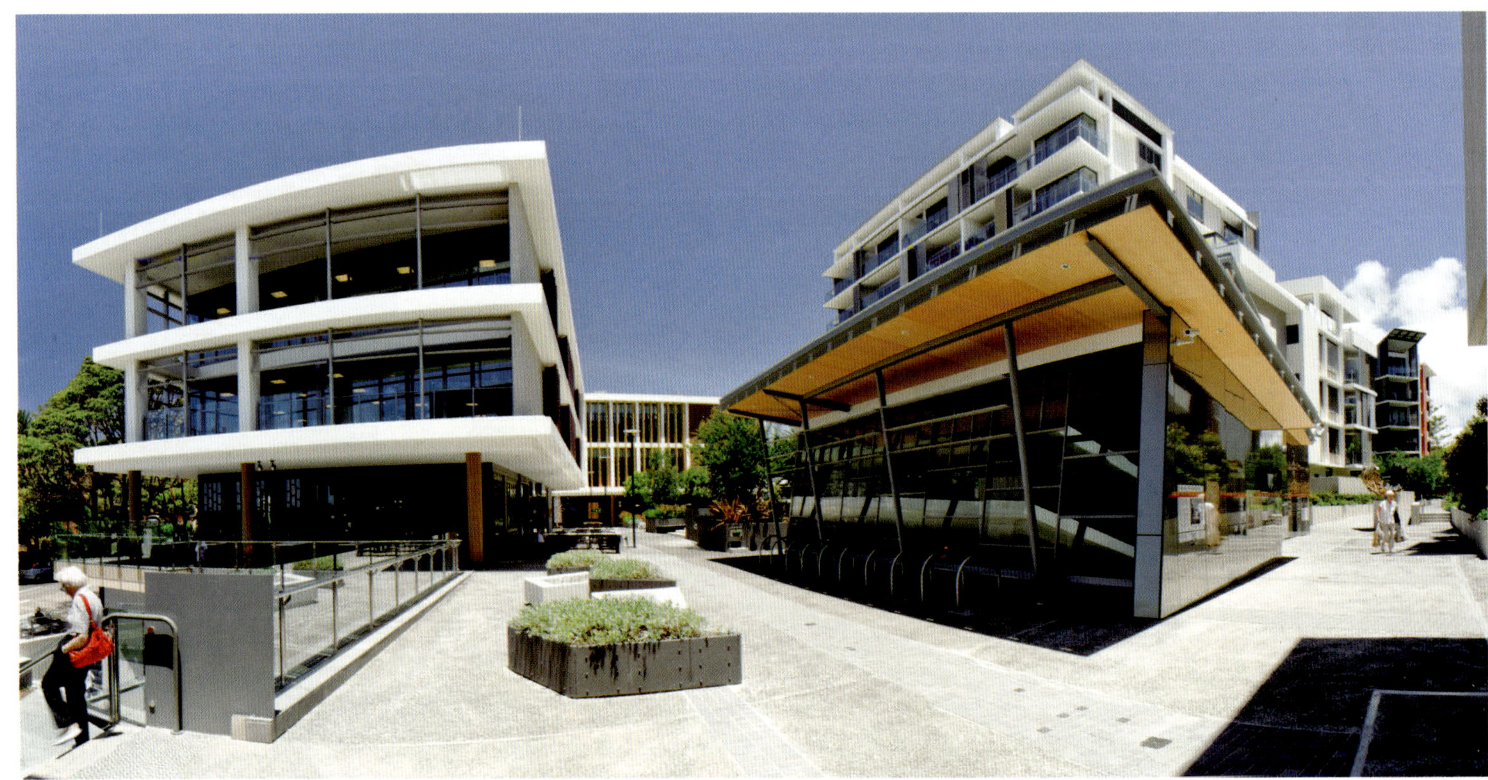

房地产广场

The Real Estate Plaza

ARCHITECT:
Avi Laiser AUArch
ARTIST:
Dana Hirsch Laiser
LOCATION:
Kaf-tet Be'November Street Bat-Yam, Israel
AREA:
550m^2
AWARD:
Project of the year by Israeli Architecture Quarterly
TYPE:
Private spaces in the public realm
CLIENT:
The city of Bat-Yam
PHOTOGRAPHER:
Hila Laiser-Beja, Asaf Evron, Orna Marton(panoramas)

建筑师：
Avi Laiser AUArch
艺术家：
Dana Hirsch Laiser
所在位置：
以色列巴特亚姆 Kaf-tet Be'November Street Bat-Yam
面积：
550m^2
所获奖项：
以色列建筑季刊年度项目
风格：
公共领域的私人空间
委托人：
巴特亚姆市
摄影师：
Hila Laiser-Beja, Asaf Evron, Orna Marton(panoramas)

We wanted to take advantage of the existing conditions of the barrier wall and the vacant land strip to create an unusual public park that allows intimate/private human activities to exist in the public domain. The project's main facade to the residential street is a see-through wall with a wooden entry gate to an outdoor room lying between the existing barrier wall and the new wall. Both are at the same height but one is obstructive and the other is inviting. Bold pink neon sign written; "the REAL estate", suggesting that the real assets of dense urban cities are outdoor public spaces. For the surface of the park, we laid a continuous fabric formed concrete "blanket" that wraps over the existing acoustic barrier wall. The continuous surface starts on a man-made horizontal landscape and changes gradually to a sloped vertical wall. In this surface we curved seven cut-out wood niches that perform as intimate private spaces in the public urban landscape. Each niche takes the form of the human body as a single, couple or a group.

我们希望利用现有的围墙条件以及空地来打造非同寻常的公园，在这片公共领域上可以进行私密的人类活动。项目面向居住区街道的立面由一堵透明的墙组成，通过木制的入口大门可进入处于现存围墙和新墙之间的室外居室。这两堵墙的高度一致，然而其中一堵墙是没有门的，而居民可以通过另一堵墙进入园区。粉红色霓虹灯上写着："房地产"，意味着高密度城市中的不动产是户外公共空间。
至于公园的表面，我们在现有隔音墙外裹上了织物形状的混凝土"毯子"。这种连续面层始于一个人造的水平景观，向倾斜的直立墙不断变化。我们还在表面建了7个木制壁龛，在这个公共都市景观中作为私密空间而存在。每个壁龛都是单独的人形，或双人形，甚至是多人形。

© Orna Marton

The project is collaboration between Avi Laiser, an architect with Dana Hirsch laiser, a performance artist. Dana has worked with teenagers form the local community center to create movement performances that happened at the site. The focus of the work was to explore individual human behavior in the private domain and introduce it to the public domain. The performances created a dialogue with the unique qualities of the built project and presented to the visitors the free essence of feeling at home outside.

这个项目是由建筑师阿维・赖萨和行为艺术家德纳・赫希・赖萨共同合作完成的。德纳跟当地社区中心的青年一起策划场地上的活动表演。这项工作的重点是在私人领域探索个人行为，并将其引进公共领域。这种性能与每个建成项目的特点形成关联，让游客就像在自己家一样自在。

花岗岩、黄铜工工艺广场
Granite, Brass Crafts Plaza

LOCATION:
Innsbruck
LANDSCAPE ARCHITECT:
Otto Kapfinger

项目地点：
因斯布鲁克
景观设计师：
Otto Kapfinger

After five years' planning and construction, the car-free zone of Maria-Theresian-Straße in Innsbruck is now completed and was opened to the public with a street party on Saturday, August 6, 2011.

位于因斯布鲁克无车的 Maria-Theresian 大街经过五年的规划和建设，现在已完成了，于 2011 年 8 月 6 日星期六向公众开放，还在街道上举行了庆祝典礼。

Converted into a pedestrian zone, the northern section of the street has already been in use for two years now and, with its choice of robust materials, has successfully withstood the daily rush of users. The granite paving now extends from the Triumph Gate to the Old Town, providing much room for urban strolls with a view of the mountains.

街道的北段被改造成一个步行区，现在已经使用两年了，因为选择使用了结实的材料，承受每天川流不息的人群是完全没问题的。花岗岩路面从凯旋门一直延伸到老城区，给游逛提供了充足空间，而且还能欣赏到山景。

A redesigning of Maria-Theresien-Straße that does justice to the significance that the street has in the townscape of Innsbruck: the goal was to create an urban site with a rich atmosphere that invites strolling, hanging out, and meeting people.
The identity of the site derives from the tension between urbanity and a panoramic view into nature, between past and future, between a specific character and a connective function in the urban structure of Innsbruck.

Maria-Theresian大街的重新设计充分体现了它在因斯布鲁克城市景观中的重要性;目标就是造就一个带有浓郁氛围的都市场所，为人们散步、闲逛和聚会提供方便。它的特色在于紧密结合了城市与开阔的自然风景，过去与未来，既带有独特性，又与因斯布鲁克整个城市布局相容。

B
U
R
G

Two defining materials, granite and brass, balance these dualities in the redesigning: a slab carpet of four different types of granite creates a coherent square surface, and a network of brass-colored ground plates with street furniture growing up from it defines the square area proper in the middle of the street.
At night, the walking zones alongside the house façades are brightly lit, while low-set lighting in the middle of the square enables a view of the mountain silhouette and the stars above.

两种核心的材料：花岗岩和黄铜，在重新设计中平衡了这些双重性：四种不同花岗岩石板铺设成一个协调一致如地毯般的广场、网状点缀的黄铜色地砖，还有矗立着的街道设施使街道中间的广场空间别具特色。
在夜晚，房屋前的步行区被照亮，而在广场低位设置的照明能让人欣赏到头顶的繁星和山峦的剪影构成的画面。

TABAK ZIGARREN

丹麦哥本哈根滨水空间广场
Kalvebod Waves

ARCHITECT:
KLAR Architects
COLLABORATION:
JDS, Niras, Sloth Moller
LOCATION:
Kalvebod Brygge, Copenhagen Harbor, Denmark
AREA:
4,000m^2
CLIENT:
Copenhagen Municipaliity Lokale- og Anlxgsfonden
PHOTOGRAPHER:
KLAR, JDS, Ursula Bach

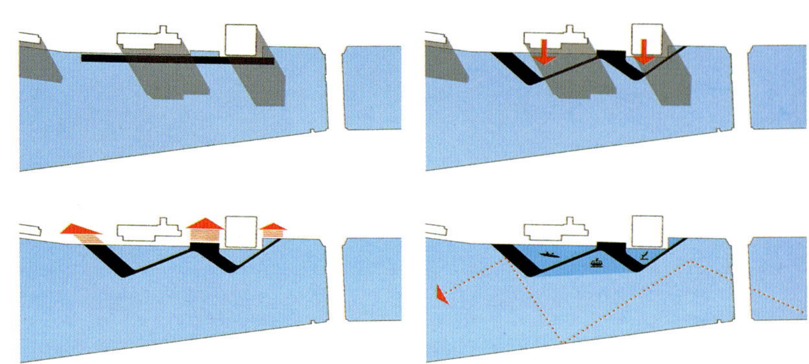

建筑设计：
KLAR Architects
合作：
JDS, Niras, Sloth Meller
项目地点：
丹麦哥本哈根港，Kalvebod Brygge
面积：
4000 平方米
项目委托：
Copenhagen Municipality Lokale- og Anlxgsfonden
摄影师：
KLAR, JDS, Ursuia Bach

The City of Copenhagen opens a new waterfront to the public; the Kalvebod Waves finally make this inner city waterfront accessible and attractive to the public. We put our focus on two major design aspects: to create urban continuity and to locate ourselves on the sunny spots of the water.

这是由DS/JULIEN DE SMEDT ARCHITECTS设计的丹麦哥本哈根Kalvebod Waves公园。该公园坐落于 Brygge 岛，Kalvebod Brygge 区，是哥本哈根夏季最受欢迎的景点之一。该岛屿是 Kalvebod Brygge 区最有发展潜力的海岛，但是可惜的是，到目前为止，仍然没有官方的监督管理和维护，属于荒岛。

What has doomed the Kalvebod area until now were the long. We studied the course of those shadows throughout the day and the year and located two main pockets of shadow-free zones. We decided to program those areas as resting islands on the water. From there on, all we needed was to find an active way to reconnect those islands to the urban network and to make them relate to the city's infrastructure.

这个新的海滨具有极大的公共活动潜力，随着中央火车站和 Tivoli 小镇的连通，哥本哈根著名的城市公园 Kalvebod Waves 将会和 Kalvebod Brygge 连成一体，将内城的活动和海港相接，体现哥本哈根现在城市滨水区的优势。该项目包括两个广场，延伸至水面上，能够让游人尽情享受阳光和海风。在南部，它可为各种活动提供公共空间设施。在过去的 10 年，哥本哈根一直在加强 Kalvebod Brygge 区的发展，以适于团体、公司、集会和大型节日活动等在海滨完成。

© KLAR Architects

The 4,000m^2 development consists of a "Waving" pier, where the citizens and visitors of Copenhagen can explore the waterfront from different levels and enjoy its amazing views. From here you can take a walk or boat-tour, rent a kayak in the kayak-hotel, get a coffee in one of the container shops, sit in the sun or simply enjoy an exciting and active public space. The "Waving" form of the development plays with varying distances to the water level, while creating a variety of basins for different water activities.

项目占地 4000 平方米，位于丹麦哥本哈根码头，为城市带来全新的海水浴场和水上运动理念。公共广场的设计理念是保持城市连贯性，并且增添水上运动设施。起伏的沿海栈道设计参照了码头突出结构的投影运动形状，使曾经黯淡沉闷的港口区域充满活力与能量。无遮阳装置的露天休息岛被木质漫步道路网连接起来。尽管历史上港口是工业活动的冷峻场地，但沿海栈道项目赋予了其新鲜的能量，开启了码头区域新的篇章。

休闲广场景观
LEISURE SQUARE LANDSCAPE

Cow Hollow 学校操场

Cow Hollow School Playground

LOCATION:
San Francisco, USA
DESIGN COMPANY:
SURFACEDESIGN INC.

项目地点：
美国 旧金山
设计公司：
SURFACEDESIGN INC.

During its short life as a public park and national landmark district, San Francisco's Presidio has become an enduring cultural and environmental resource for the city. Strict stewardship of park is overseen by the Presidio Trust whose mission is to both protect and restore the resources of the park and to provide opportunities to residents, visitors and businesses in the Presidio access to this unique place.

旧金山普雷西迪奥在她作为一个公共公园和国家地标区的短暂历史中，已经成为这座城市的一个永久的文化和环境资源。普雷西迪奥信托公司负责对该公园严格管理，保护和维护公园的资源，为当地居民、游客和商人提供机会来享受这个独一无二的人间乐园。

When the Cow Hollow School, a preschool for children 2-5 years of age, decided on a new home to expand its facilities and educational mission, the Presidio was a logical choice as the combination of nature and history coincided with the specific goals of its curriculum. Located a short walk from the Presidio's historic Parade Ground, the school contacted the landscape architect to conceive of a playground design that is an extension of the classroom. Extensive workshops with school teachers, administrators and parents helped the designers to understand the school's education mission: learning through play; exploration and discovery; and nurturing relationships between children, parents and teachers.

Cow Hollow 学校（一所招收 2~5 岁儿童的幼儿园）决定建一所新校舍，从而扩大设施和增添教学功能。普雷西迪奥的自然与历史相互结合，这一点符合学校课程的特殊目的和要求，是一个复合逻辑的理性选择。学校位于具有历史意义的普雷西迪奥阅兵场附近，步行几分钟就可以到达。学校向园林设计师表达了自己的想法：把操场作为教室的延伸。园林设计师与学校的教师、管理者和学生家长充分讨论，他们帮助园林设计师去理解和解读学校教育的使命：在玩中学，探索与发现，呵护学生、家长和老师的关系。

Curriculum at the school is based on the Reggio Emilia approach, a philosophy and educational program that teaches principles of respect, responsibility and community by encouraging children to be their own teachers and by providing a supportive and enriching environment based on the interests of the children. As the organization of the physical environment is linked to Reggio Emilia's early childhood program (often seen as the "third teacher"), the design of the playground was seen as critical to the curriculum of the school.

学校的课程是基于瑞吉欧·艾米利亚的教育方法，她提出了一种理念和教育方案：教学的原则在于通过鼓励孩子自学的方式，根据孩子们的兴趣来提供丰富的辅助环境，让学生学会尊重、责任和交流。由于自然环境的安排状况与瑞吉欧·艾米利亚的学前教育计划（环境被看作是第三个老师）紧密联系，因此，操场的设计被认为是学校课程设计的关键。

The landscape design for the Cow Hollow School is indebted to nearby physical surroundings of the San Francisco Bay area and to the natural setting of the Presidio. References to the local landscape typologies of beaches, forests, the Marin headlands and bay tidal wetlands were incorporated into the physical configuration of the school yard. The playground design incorporates the educational and interactive requirements of the Reggio Emilia program by emphasizing the landscape as an extension of the classroom, by providing a variety exterior spaces for self-instruction, and by emphasizing direct visual and physical connections between the classroom and natural landscape.

学校景观的设计得益于附近旧金山湾地区周围的自然环境和普雷西迪奥的自然环境。我们在布局学校操场时参考了当地的沙滩、森林、马林陆岬和海湾潮汐湿地等典型的景观，该设计考虑了瑞吉欧·艾米利亚的教育方法中对教育和互动提出的要求，强调风景是教室的延伸，强调教室和自然景观之间直接的感观联系，为学生提供一个多样化的自我学习的外部空间。

Getting building approvals to build the new playground from the Presidio Trust proved to be a formidable task for the client and the designers. The playground design needed to demonstrate that there would be no impact on the native plant habitats at the site. Historic and cultural resources below the surface of the site needed to be protected as well. All built and planted elements of the playground would have to be completely removable as to not disturb the ancient foundations of the original Spanish Presidio, the namesake of the park. As a result, trees and plant materials were placed away from the Presidio foundations and materials such as rammed earth and path fines were selected as not to impact archeology below.

从普雷西迪奥信托公司得到的建设这个新操场的许可表明了此次设计对于客户和设计者来说是一个艰巨而伟大的任务。操场的设计要体现出对当地的植被不会有影响。该场址地下的历史和文化资源也需要保护。操场上所有建造的和种植的东西必须是可完全移除的，为的是不破坏西班牙普雷西迪奥公园原始的古代地基。所以，树木和其他植物被安置在远离雷西迪奥古代地基的地方，建筑材料如夯土和路面细料都经过精挑细选以不影响下面的古文物。

乡村别墅公园和散步广场

Landhauspark and Promenade

LOCATION:
Linz, Austria

AREA:
20, 000 sqm

DESIGNER AND COUNTRY OF DESIGNER :
el:ch landscape architects, Germany

PHOTOGRAPHER:
Christian Henke, Alexander Henke

AWARD:
Österreichischer Bauherrenpreis 2009
(Austrian Client's Price)

DESIGN COMPANY:
el:ch landscape architects

项目地点：
奥地利 林茨

面积：
20 000 平方米

设计师及设计师所在国家：
el:ch landscape architects, Germany

摄影师：
el:ch landscape architects, Germany

奖项：
Österreichischer Bauherrenpreis 2009
(Austrian Client' s Price)

设计公司：
el:ch landscape architects

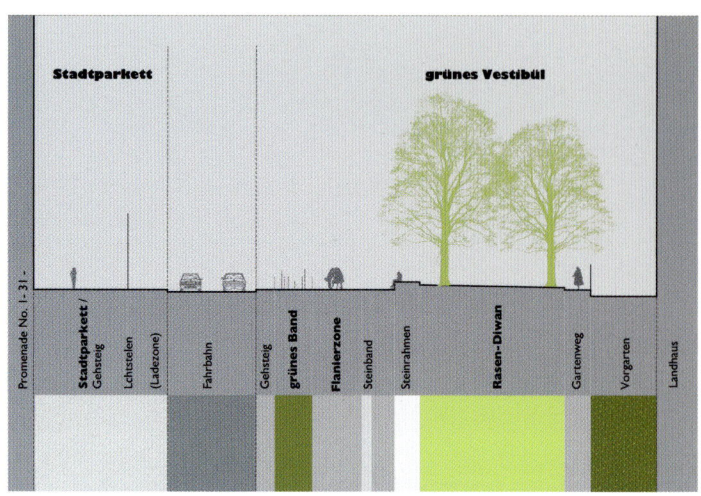

The site in its present form came into existence only in 1800 when a fire destroyed most of the city. The rampart was razed and transformed into a horticultural designed area. The 20th century's developments in transport had a negative impact on the space, making it primarily a passageway for motorized traffic. Pedestrian movement was limited to narrow areas, close to the buildings on both sides, available space devoured by parked vehicles and an unimpeded clutter of urban furniture.

项目建筑场址的外观布局在 1800 年才形成，当时整个城市的大部分被一场大火毁掉。城池的围墙被夷为平地，然后这块地方进行了园艺设计。二十世纪的交通发展对该空间起了负面作用，它基本上变成了一条机动车通道。步行只限于一些狭窄的地方，挨着两侧的楼，这里还停放着车辆和摆放着一堆杂乱的家具。

In 2005, the government of Upper Austria and the Linz city council agreed in a joint effort to install an underground car park and remodel the area's surface. The resulting competition brief required entrants to develop a discernible identity for the site, providing usable space for a variety of user groups, while retaining all the existing trees, which are scattered without recognizable pattern all over the area.

2005 年，上奥地利州政府和林茨市委员会同意共同努力修建一个地下停车场，对整个地方的外部进行改造。随之而来的设计竞赛纲要要求参赛者为该场地打造出一种鲜明的特色，给各类用户群体提供可用的空间，但对散布在整个园区那些无序排列的树要予以保留。

After our preliminary site visit, we soon realized: the space's identity was already there. It had just become unrecognizable beneath layers of confusing additions and competing requirements. We observed a sophisticated scenic urban facade along the southern and western side of the site. The northern and eastern edges of the L-shaped plot were adorned with an amazingly pictorial population of trees. This bipolar identity, oscillating between the urban side of the actual Promenade and its green counterpart – the Landhauspark, was the atmospheric concept we strove to make comprehensible for the users.

初步实地考察之后，我们很快意识到：这块空间的特征就已经在那儿了，只是在设计上层层眼花缭乱的补充添加和比赛要求的掩盖下变得模糊了。在这块 L 形场地的南边和西边我们发现了浓重的都市景观，在北侧和东侧边缘栽着许多迷人如画的树。一边是散步广场的都市气息，一边是绿色公园的乡村格调，这种双重风格正是我们努力向用户们所传达的意境。

Both sides draw from the other's complementing qualities: The Park gives a green vestibule to Promenade and Landhaus, connecting with the close-by Castle Mountain. The Promenade stretches its "urban parquet" towards the city's main street along its unique architectural backdrop.

两边的特色互为补充：公园与附近的城堡山相连，给散步广场和乡村别墅提供了绿色入口；散步广场像一块铺设地板的城市道路向城市的主街延伸，街边是独一无二的建筑。

适合居住的户外空间广场
Geografia Habitable Outdoors Space

ARCHITECTURE & DESIGN:
Hector Ruiz-Velazquez MArch
LOCATION:
Valencia, Spain
CLIENT:
FERIA HABITAD VALENCIA
LIGHTING:
Vondom
POOL:
Hector Ruiz-Velazquez MArch
INTERNAL WALL:
Acrylic fabric
PHOTOGRAPHER:
Pedro Martinez

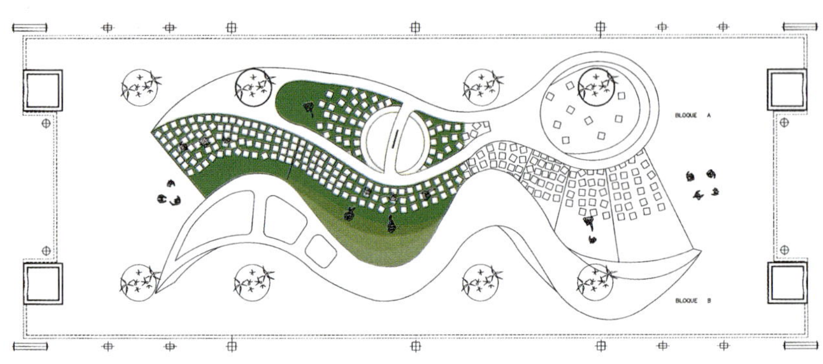

建筑师和设计师：
海克特鲁尔兹一委拉斯开兹
项目地点：
西班牙巴伦西亚
委托人：
瓦伦西亚展览公司
照明：
伏尔德姆
水池：
海克特鲁尔兹一委拉斯开兹
内墙：
丙烯酸构造
摄影师：
佩德罗·马丁内斯

Emerge in this space of lights and shades is to enter an allegory to the Spanish geography, to the Spanish dynamic live, to his ideas and his passions. This dynamic structure is a totally interior surrounding that connects with the exterior thanks to dynamic forms that evoke the nature of a Made Spain rich in contrasts. This space was designed to wake up the senses up the visitors through the vision in the form, the smell with the real grass in the floor that give the olfactory sensation of being outdoors, the tact by touching the real nature of the curve walls that follow all the trail, and the hearing by the cascade in the heart of the space. It was built in ten days, all done in a metallic curve structure covered with Acrylic fabric.

这片光与影的空间展示了西班牙的地形、西班牙多姿多彩的生活及其理念与情感。这种动态结构完全是一种室内外环境的结合体，其动态形式唤起了充满对比的西班牙自然本质。这片空间的目的是为游客提供各种感官体验，地面上真实的小片草坪能给予我们户外的嗅觉体验，游客通过触摸弯曲的墙面能感受真实的大自然，而在空间的中心还能听到小瀑布的声音。其结构为金属曲线，表层覆盖丙烯酸构造，于10日内全部完工。

This space offers a symbolic vision associated with the soil, with the land, with the original and authentic. Ramps and elevations in real balance of the unlimited curves.

在这里，人们可以感受到这块土地最原始、最真实的土壤。斜坡和立面之间达成了完美的平衡。

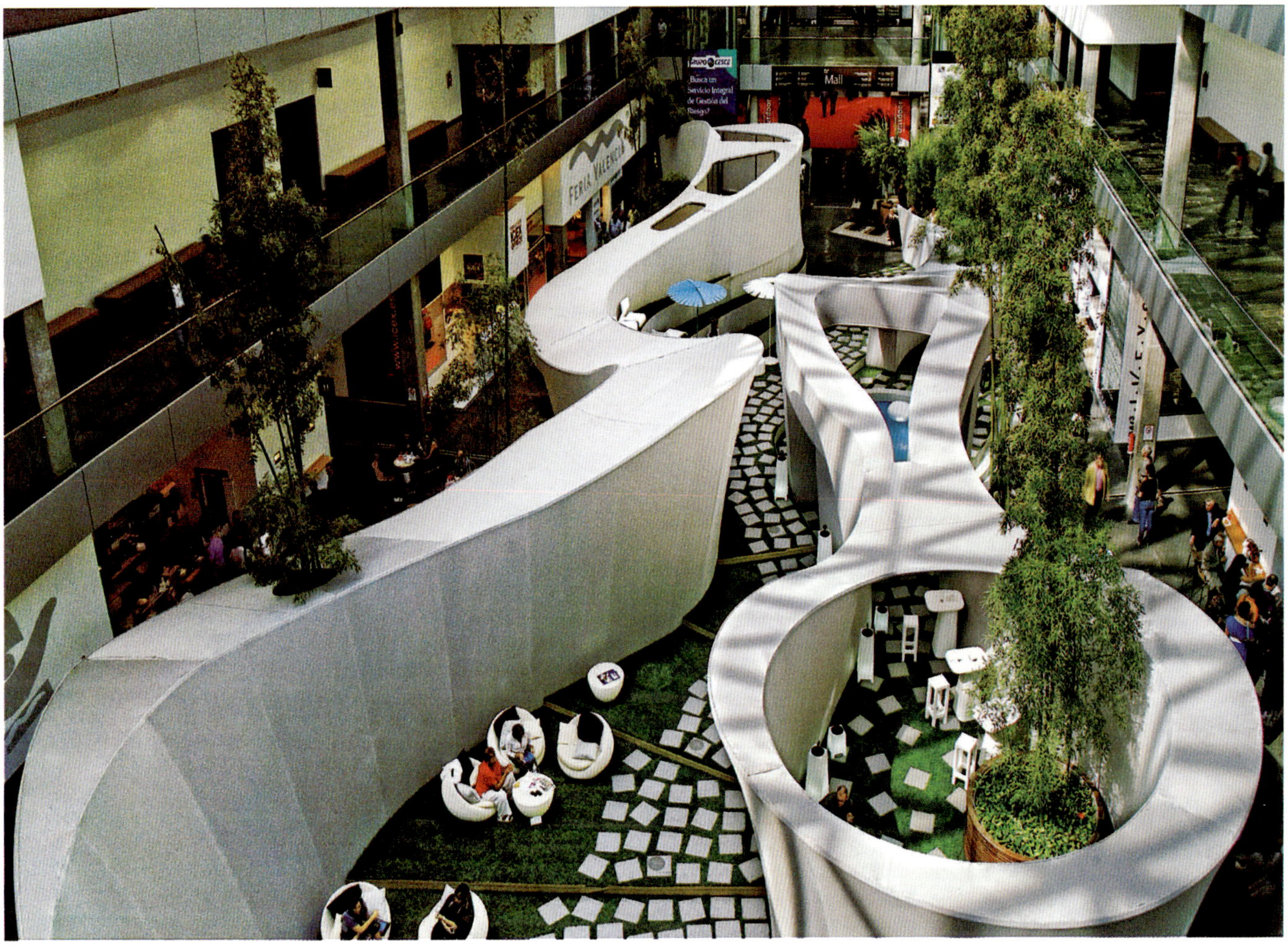
FERIA VALENCIA
Mall

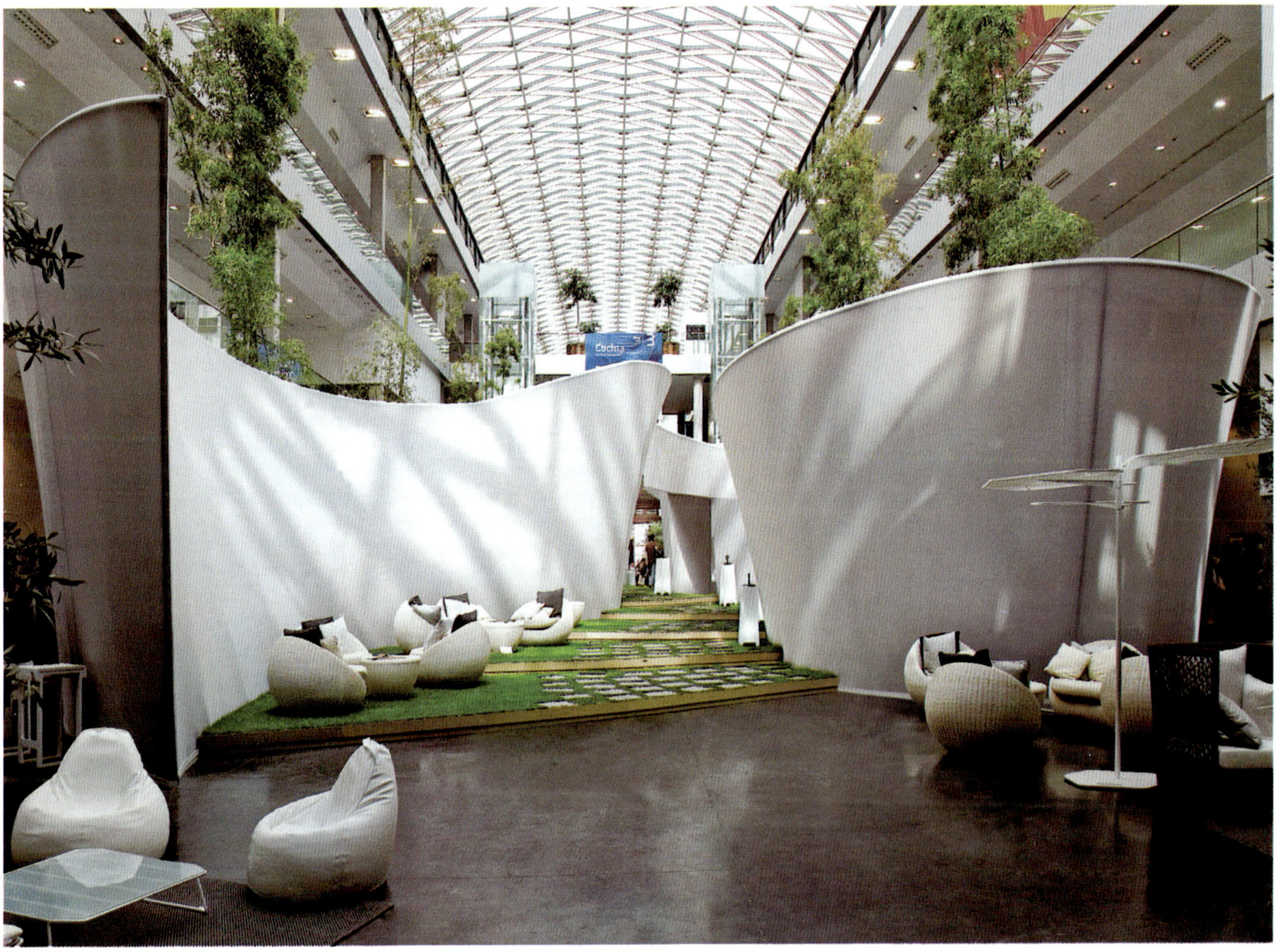
Cocina

Well-deserved honoring where the manufacturers expose their elaborated products for outdoors: a careful selection to take ourselves to the pleasure and enjoyment of our landscapes and our climate. Products that are elaborated based in innovation, sensitivity and a competitive vision, which are nourished from our way of life, always creative and searching. The Spanish temperament, so admired internationally, is opened and measures up every day in the exterior. From these open roots to the landscape, the Spanish producers offer us furniture impregnated with character.

厂商展示了他们精心制作的户外产品：一系列精选品使我们快乐地欣赏着我们的风景与气候。所展示的产品极具创新性、灵敏度及竞争力，这些灵感均源自于我们的生活方式，充满创意和活力。这里有着国际化的西班牙气息，每天的环境都符合户外的变化。从这些露在外面的树根到各个景观，西班牙的设计师为我们提供了独特的家具。

The hall of the fair open itself to welcome everyone with a new structure that represents fresh ideas and the Spanish passions, a place where to walk and to observe every detail.

大厅一直处于开放状态，以其崭新的结构欢迎每一个造访者，这个结构代表着新鲜的思维以及西班牙的热情，这是一个用来散步和观察事物的好地方。

Mall

Mall

柏林 Kladow 区 Imchen 广场和海滨长廊

Imchen Square Berlin-Kladow Redesign of Square and Waterfront Promenade

LOCATION:
Berlin, Germany
AREA:
10,000 m²
DESIGNERS:
glasser and dagenbach, landschaftsarchitekten
Sabrina Schroeder, Silvia; Udo Dagenbach
PHOTOGRAPHER:
Udo Dagenbach
DESIGN COMPANY :
glasser and dagenbach GbR
garden and landcsape architects

项目地点：
德国柏林
面积：
10 000 平方米
设计师：
glasser and dagenbach, landschaftsarchitekten,
Sabrina Schroeder, Silvia; Udo Dagenbach
摄影师：
Udo Dagenbach
设计公司：
glasser and dagenbach GbR
garden and landcsape architects

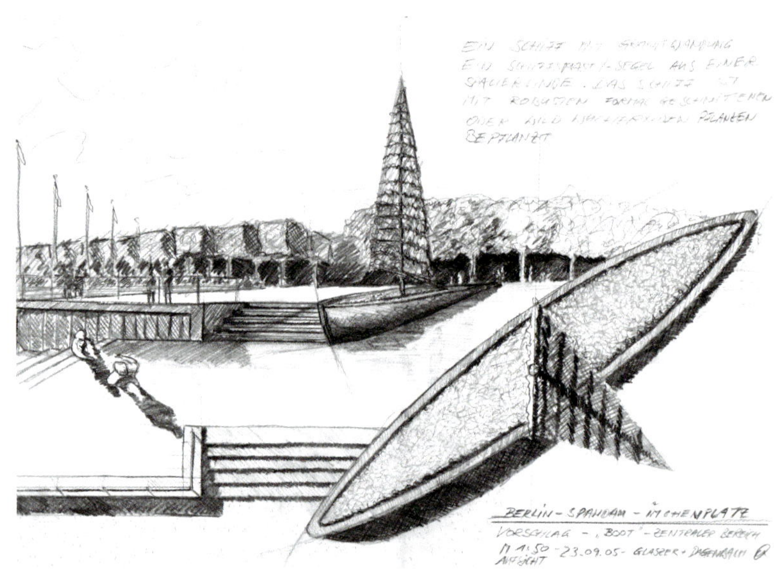

Imchenplatz Berlin-Spandau – modernizing a square and shipping pier at the river Havel in Berlin.

柏林 Spandau Imchen 广场——柏林哈沃尔河边现代化的广场及船运码头。

At the natural and soft coastal western side of the river Havel in Berlin-Spandau a very traditional square and shipping pier is located in the district Kladow: the so called Imchenplatz. Left over structures from the early sixties and an overaged Tilia alley gave a poor impression of this very important local recreation area for Berlin's inhabitants.

在柏林 Spandau 哈沃河天然柔美的西岸，一个非常传统的广场及码头位于 Kladow 区：就是所谓的 Imchen 广场。对于柏林的居民来说，20 世纪 60 年代早期留下的建筑和一个古老的椴树胡同让这个当地重要的娱乐中心给人留下糟糕的印象。

The city council of Berlin-Spandau gave us the contract to do a strong redesign of coastline alley and square.

柏林市 Spandau 委员会给我们这个合同，要求我们重新设计岸边的广场和胡同。

Further the square should be prepared for a stronger use in future as Christmas market and place for summer party events.

进一步说，这个广场作为圣诞市场和夏天举办娱乐活动的地方应该准备好在未来发挥更强的功能。

Also the old boat loading ramp should be renovated and integrated in the new design.

而且，那个古老的船装载坡道应该翻新和修复，融入新的设计当中。

We decided to create a very reduced design which integrates both – the fixed budget of 800,000 EUR and the multitasking requirements of many users.

我们决定创造一个简约的设计，用 800 000 欧元的固定预算满足许多使用者的多重要求。

Granite steps and platform follow the coastline. A canvas shaped big lawn area flows to the river.
Smaller canvas shaped areas with blood beach hedges and benches are adjusted to the walkway along the lawn and provide a good view to the river.

花岗岩石阶和平台沿着河岸边，如画布般的大草坪延伸到河边。
海岸边的树篱围成了面积较小的风帆形状地块，草坪两边放置长凳，从而绘出一幅美丽的河滨风景图。

An over 40 m long curved plantbed for a prairie style perennial plantation is situated on one side of the lawn. Plantations of Gras like Miscanthus and Pennisetum frame the red leafed hedges.

草坪的一侧是一个超过 40 米长的弯曲的花坛，里面生长着大草原多年生草本植物。

To define the big space around the boat loading ramp we placed a boat made of blank shaped basalt stone blocks. The boat is planted with lavendar.

为了装饰船装载坡道周围硕大的空间，我们放置了一个用毛坯玄武岩石块做成的小船，小船里种植了　衣草。

The alley along the river was replanted with Tilia cordata.

沿着河边的小路移植了欧洲小叶椴。

国家展会（ULAP）广场
ULAP-Square

LOCATION:
Berlin, Germany
AREA:
1.3 ha
PHOTOGRAPHER:
Rehwaldt LA, Dresden
DESIGN COMPANY:
Rehwaldt LA, Dresden

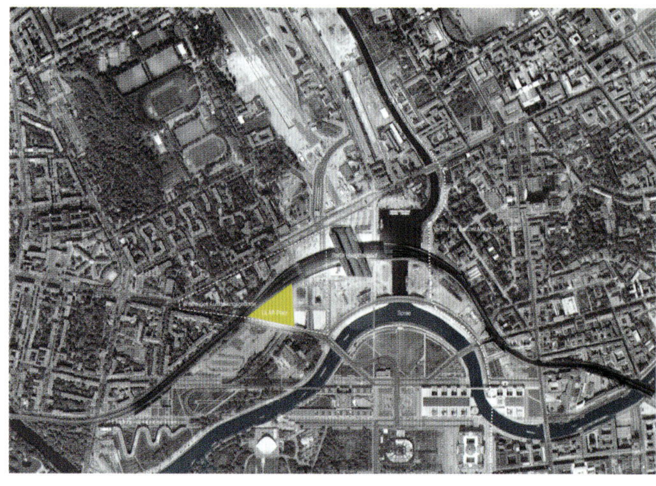

项目地点：
德国柏林
面积：
1.3 公顷
摄影师：
Rehwaldt LA, Dresden
设计公司：
Rehwaldt LA, Dresden

In 1879 an exhibition area was established west of the ancient Berlin between Lehrter Bahnhof, Alt-Moabit and Invalidenstraße. A site's characteristic was the division into two areas by the still existing commuter railway viaduct.

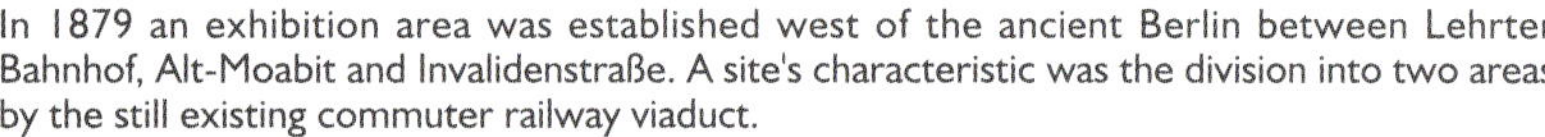

1879 年，在柏林西部柏林中央火车站、Alt-Moabit 和 Invalidenstraße 之间建立了一个展览区。展览区最显著的特征就是被一座通勤铁路高架桥一分为二。这座铁路高架桥至今仍保留着。

Although planned as temporary arrangement for an industrial exhibition the area was shortly developing to a permanent showground in Berlin named Universum-Landes-Ausstellungs-Park, short ULAP (Universe-National-Exhibition-Park). Until short time before World War I most of the big exhibitions for trade, hygiene, art and technics took place here.

这个展览区最初是作为一个临时工业展览区进行规划的，但是很快就发展成为柏林市一个固定的展览会场，并命名为国家展览会场，简称 ULAP。直到第一次世界大战前不久，大部分大型的商品交易、卫生、艺术和技术展览会都在这里举行。

The first show was the trade exhibition in 1879, which was because of its innovative products the precursor for subsequent big industrial exhibitions. It also marked Berlin's beginning as great exhibition city.

第一次展览是 1879 年举行的商品交易展览会，会上新颖的产品为以后大型工业展出奠定了基础，从那以后，柏林逐渐发展成为世界重要的展览场所。

For this reason an exhibition building was erected on both sides of the railway viaduct, which also used the viaduct archs as showground. The railway line was not operating by then. The planner of the site had to deal with a 4 m height difference to the adjacent street level.

为此，铁路高架桥的两侧各建起了一座展览大楼，高架桥的拱形结构也被用作展览场地。那时这条铁路线还没有开通。场地的规划者需要解决场地与附近街道之间的 4 米高差的问题。

At that time the main entrance of the exhibition area was situated on the street Alt-Moabit. Via a wide two-way staircase featuring a water cascade the visitors attained the main building.

那时，展览区的主入口位于 Alt-Moabit 大街上，游客可以通过一部宽阔的双向楼梯来到主建筑，小瀑布成为楼梯的亮点。

From there people attained a forecourt with water basin and flower beds which also integrated the main staircase. The site west of the railway viaduct was landscaped. Besides the single pavilions and facility buildings a large artificial lake formed the center of the fairground.

从这里，人们可以进入一个前院，前院里面有水池和花坛，与主楼梯融为一体。铁路高架桥的西侧进行了景观美化。凉亭与设施楼之间有一个大型的人工湖，成为会场的中心。

After the devastating fire of the exhibition hall shortly before the opening of the General German Exhibition for hygiene and rescue in 1882 an intensive discussion about the future of ULAP emerged. Under the pressure of an expected universal exhibition in Berlin possible alternative fairgrounds were discussed. Even the usage of Lehrter Bahnhof was considered which was planned to shut down by that time.

在一场毁灭性的火灾之后，1882 年举办德国卫生和救援展会之前不久，人们对 ULAP 的将来进行了大量的讨论。由于柏林将要举行一次世界展览，因此有人提议在这里建一个交易会场，甚至考虑到了使用当时即将关闭的柏林中央火车站。

伊尔桑花园广场

Ilsan Xi WICITY

ARCHITECT:
Group Han Associate
LANDSCAPE:
Construction CORYO landscape architecture
CONSTRUCTION:
GS E&C
LOCATION:
A1, A2, A4 Block, Siksa-Housing development district, Siksa-dong, Ilsandong-gu, Goyang-si, Gyeonggi-do
SITE AREA:
339,591 m^2
LANDSCAPE AREA:
157,416m^2
NUMBER OF HOUSEHOLDS:
4,683
AWARD:
President's award in IFLA APR Award for Landscape Architecture 2011, Grand Prize in The 1st Green Space Award, Grand Prize in Korea Color

建筑设计：
Group Han Associate
园建工程：
CORYO landscape architecture
结构设计：
GS E&C
项目地点：
A1, A2, A4 Biock, Siksa-Housing development district, Siksa-long, Ilsandong-gu, Goyang-si, Gyeonggi-do Site
场地面积：
339 591 平方米
景观面积：
157 416 平方米
户数：
4683
获奖情况：
President's award in IFLA APR Award for Landscape Architecture 2011, Grand Prize in The 1st Green Space Award, Grand Prize in Korea Color

Bubble Garden

Rhythmical Field

Themed to "Water Garden" that links the existing streams in the forest, Compound I includes various water features including Eco Stream and Lake Plaza. Especially the "Lake Plaza" provides the residents with a multipurpose space that can be used for water play in summer and events in spring and fall. The "Bubble Garden" is composed of five theme parks that can be used for family get-togethers and include sculptures variously themed to wood, animals and stone. Created by exploiting the topographical characteristics that make different levels on the compound, the "Eco Stream" include rocks, soils and diverse wild flowers to provide a habitat for speaes.

该设计以水景花园为主题并且场地内部的水与森林中的溪流相连接，一号场地包含着水的不同形态比如生态溪流和湖心广场部分。特别是湖心广场能提供多用途空间，比如夏季戏水，春季和冬季的重大活动等。沸腾园由五个主题园组成，包含动植物及石头的雕塑，可用于家庭的团聚活动。生态溪流在设计时充分利用场地内不同层次上的地形特点，包含假山石、土壤以及为生物提供栖息地的不同野花草。

Compound 2 has the beautiful and natural-looking "Forest Garden" that highlights the green preserved in the center. A picturesque waterfall that takes its motive from "Cheonseondae" in Mount Gumkang in DPRK is reproduced to create beautiful natural scene with Korean traits in it, while the compound is completed by tapping into the knowhow from Massimo Venturi Ferriolo, the world-class Romantic landscapist and EDAW, a world-class landscape and urban design company. The "Urban Jungle" as a resting place that encapsulates a harmony between man and nature with the motive of a forest was designed in cooperation with Prof. Ferriolo. Besides the "Urban Jungle" as a contemporary interpretation of forest, the "Garden under the Petals" that designs trees as contemporary sculptures, and the "Pine Tree Landscape" that excellently matches the contemporary building cut a figure.

二号场地有一个美丽自然的森林花园，突破了中心被保护的绿地，设计灵感来自朝鲜 Gumkang 山风景绝佳的瀑布，还有点韩国的风貌。项目模仿和学习了世界级的景观和城市设计公司易道的设计。都市丛林与 Ferriol 教授合作设计，目的是创造一个人与自然和谐共存的休憩环境。另外都市丛林与周围的森林相协调，花瓣下花园将树木作为雕塑，松林景观与周围建筑风格相匹配。

2009 什未林国家花园广场

National Garden Show Schwerin 2009

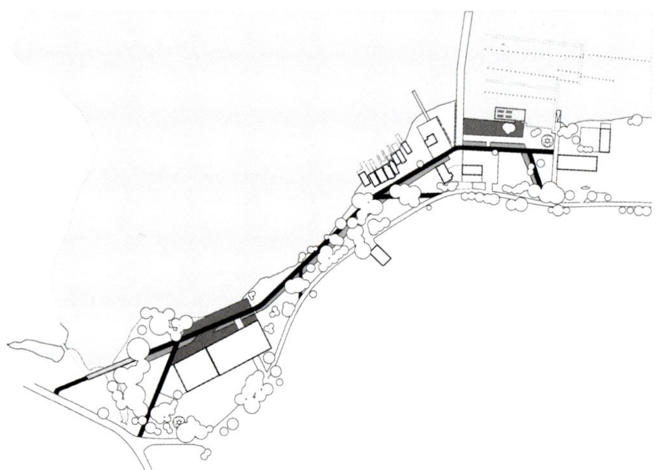

LOCATION:
Schwerin ,German
AREA:
3.1 ha
DESIGN COMPANY:
GESELLSCHAFT VON TOPOTEK I LANDSCHAFTSARCHITEKTEN MBH

项目地点：
德国什未林
面积：
3.1 公顷
设计公司：
GESELLSCHAFT VON TOPOTEK 1 LANDSCHAFTSARCHITEKTEN MBH

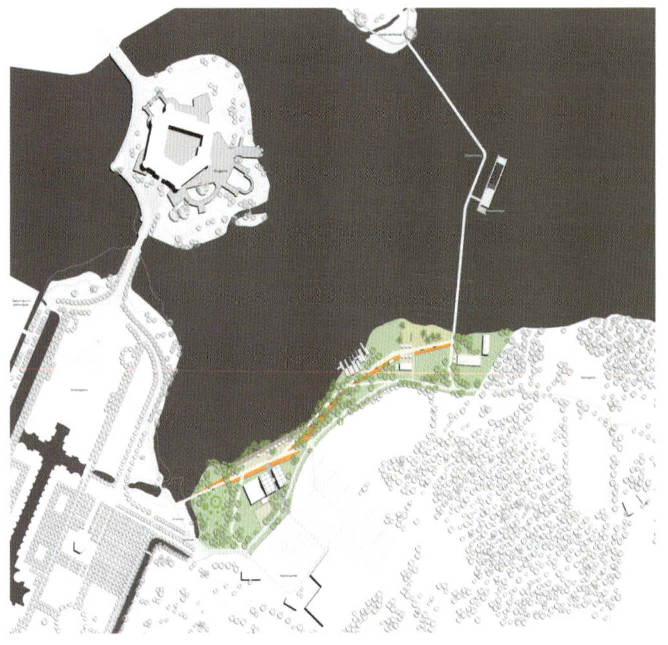

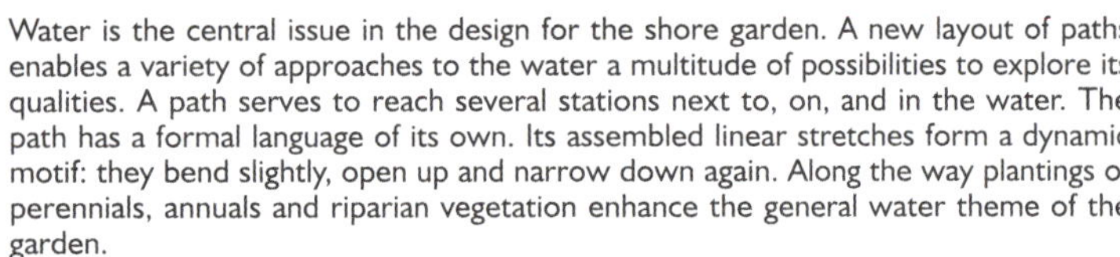

Water is the central issue in the design for the shore garden. A new layout of paths enables a variety of approaches to the water a multitude of possibilities to explore its qualities. A path serves to reach several stations next to, on, and in the water. The path has a formal language of its own. Its assembled linear stretches form a dynamic motif: they bend slightly, open up and narrow down again. Along the way plantings of perennials, annuals and riparian vegetation enhance the general water theme of the garden.

水是海滨花园设计的中心问题。小路新的布局以各种方式探索水的潜质，通向水边、水上和水中的各个站台，每条小路都有自己正式的语言，它们组合在一起，呈线性延伸，形成了一个动态的主题：时而弯曲、时而开阔、时而狭窄。沿着小路生长着一些多年生、一年生植物，突出了花园水的主题。

Visitors reach the shore garden coming from the castle. The path along the waterfront takes them across an estuary of the lake, through reeds and a glade with a playground – all the way to the terrace of the rowing club. Here, overseeing picturesque water-lily fields the terrace affords a panoramic view across the lake and back towards the castle. The path then follows the natural shore for a while, under a shady tree canopy, and finally reaches a temporary beach where white sand and palm trees create a surreal piece of maritime landscape. From the beach the path swerves onto the lake and leads to the other side of the lake on a pontoon construction. In the middle of this footbridge a floating water-lounge is moored, accentuating and structuring the stretch of way across the lake. Here, at the bar one can have a cocktail or a cool, refreshing glass of water.

游客可以从城堡来到海滨花园。水边的小路带领着游客穿过湖港、芦苇和一片有运动场的空地 —— 一路来到划船俱乐部的平台上，从平台上可以俯瞰如诗如画的睡莲，一览湖面和后面城堡的全景。然后，小路沿着自然海岸延伸，经过一棵可以乘凉的大树，最终来到临时海滩，海滩上的白砂和棕榈树形成了一幅梦幻般的海景。从这里开始，小路转向湖面，通过一个浮桥通往湖的另一侧。在浮桥的中间有一个漂浮着的水上休息室，突出了湖上通道的伸展，在这里，人们可以在酒吧品尝鸡尾酒，或者喝一杯清凉的水。

Beukplein 广场

Square Beukplein

LOCATION:
Hague, The Netherlands
AREA:
700 m^2
DESIGNER/ARCHITECT:
Carve took care of designing, engineering, site management and surveying
TEAM:
Elger Blitz, Mark van der Eng, Jasper van der Schaaf, Milan van der Storm
PHOTOGRAPHER:
Carve

项目地点：
荷兰 海牙
面积：
700 平方米
设计师 / 建筑师：
Carve took care of designing, engineering, site management and surveying
团队：
Elger Blitz, Mark van der Eng, Jasper van der Schaaf, Milan van der Storm
摄影师：
Carve

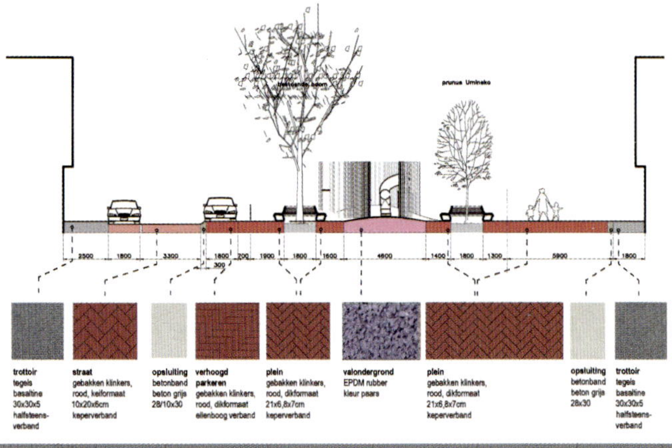

Together with local residents, Carve redesigned the Beukplein ("Beech-Square") in The Hague. By closing the street at the long side of this triangular square, the square has grown bigger and has become more cosy at the same time. The playing object with climbing, swing and slide functions is situated in a maze of coated steel poles, on an undulating purple surface.

和当地的居民一起，Carve 重新设计了位于海牙的 Beukplein 广场（山毛榉广场）。这个三角形广场长边一侧的街道被封死，使广场变大了，同时变得更加温馨。带有攀爬、秋千和滑梯功能的玩耍设施被安放在一个涂层钢柱组成的迷宫里，位于一个起伏的紫色平面上。

Boulders have been placed as informal sitting and playing elements. As a finishing touch, trees and additional furniture were placed. The trees are in bloom three out of four seasons; because of this, the square is framed by an abundance of colours throughout the longest possible time of the year. All trees are surrounded by seating to create more than enough space for all locals to hang out and enjoy their new square

安放的巨石可供坐着休息和玩耍。树和其他一些放置的设施是最后的润色。这些树四季当中有三个季节开着花；正因为此，在一年中很长的时间里，广场都被绚丽的色彩所围绕。所有的树都环绕着座椅，创造出足够的空间供所有的当地人闲逛和享受他们的新广场。

St. Pancras 教堂花园

St. Pancras Church Garden

ARCHITECT:
Studio Weave
LOCATION:
Pancras Lane, London
AREA:
190m²
PHOTOGRAPHER:
Studio Weave

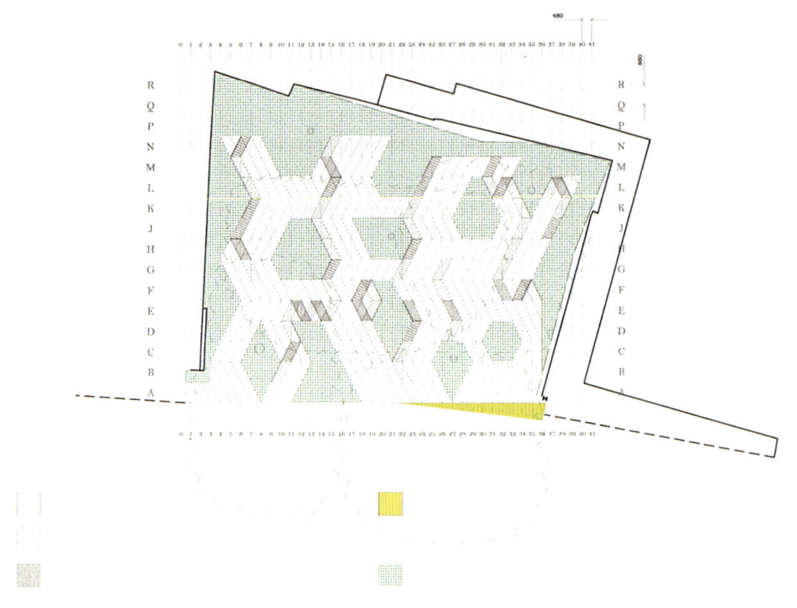

建筑设计：
Studio Weave
项目地点：
英国伦敦潘克拉斯巷
面积：
190 平方米
摄影师：
Studio Weave

Studio Weave has transformed the site of a burnt-down church in the heart of London into a unique garden filled with fantastical creatures. St. Pancras Church Garden opens up a space for public use that has remained hidden away from view for almost 350 years.

Weave 工作室已经把这个位于伦敦腹地的被烧毁掉的教堂的场地变成一个充满丰富有趣生物的独特花园。该花园也为公众提供了被隐藏了 350 余年的本可利用的开放空间。

Arranged in an irregular herringbone pattern, stone paving and carved wood weave around diamond-shaped plant beds and existing mature trees, with the wood elements rising to become seats. Each of the eight unique designs depicts patterns and animals that take their cue from the carvings of surviving Romanesque churches, with a contemporary twist. Crucial to the realisation of this project was the creative input from the Historic Carving department at City&Guilds Art School in south London, which was commissioned to handcraft the furniture from oak.

铺地石材和雕刻的木头被以不规则交叉方式环绕和编织在钻石形状种植池和现存树木周围，木构件成为坐凳。八个带有动物形象的独特浮雕暗示着罗马式教堂有幸存下来的雕塑，带有与之同时期的风格。这些批判现实主义的作品是来自伦敦城市行业协会艺术学校历史雕刻专业的创意，橡木家具也是委托他们以手工艺制做完成的。

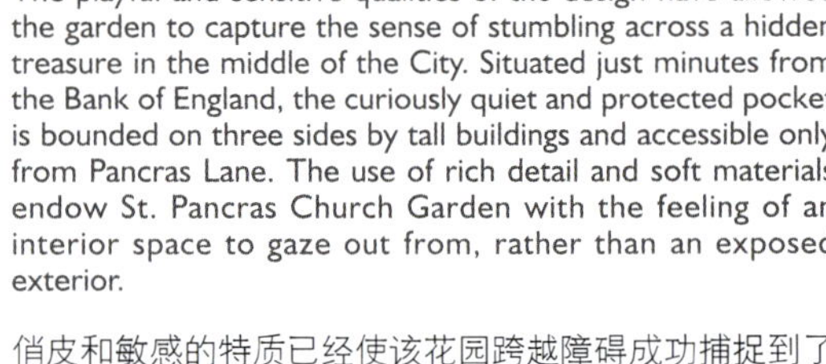

The playful and sensitive qualities of the design have allowed the garden to capture the sense of stumbling across a hidden treasure in the middle of the City. Situated just minutes from the Bank of England, the curiously quiet and protected pocket is bounded on three sides by tall buildings and accessible only from Pancras Lane. The use of rich detail and soft materials endow St. Pancras Church Garden with the feeling of an interior space to gaze out from, rather than an exposed exterior.

俏皮和敏感的特质已经使该花园跨越障碍成功捕捉到了这个城市中部隐藏的财富。该地距离英国央行只有几分钟的路程，三面都有较高的建筑围合，仅从潘克拉斯巷有个入口。丰富的细节和软质材料的使用使得花园让人觉得更内向而不是很张扬地暴露在外。

Ecole del Rusco 主题活动广场

Circolare, Ecole Del Rusco

LOCATION:
Bologna, Italy
AREA:
4500 sqm (Liber Paradisus Square)
DESIGNERS:
Ciclostile Architettura
PHOTOS:
Giacomo Beccari, Ciclostile Architettura

项目地点：
意大利博洛尼亚
面积：
4500 平方米（Liber Paradisus 广场）
设计师：
Ciclostile Architettura
照片：
Giacomo Beccari, Ciclostile Architettura

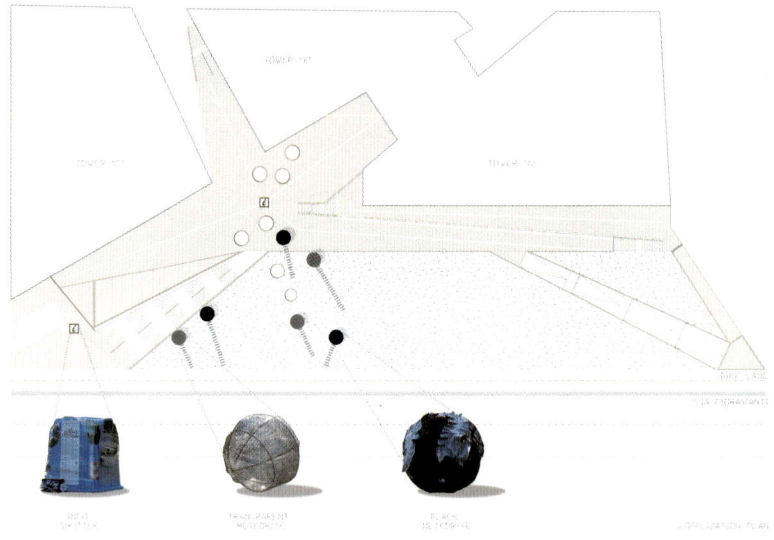

For the fourth edition of Ecole del Rusco: the exhibition about art and recycling in Bologna, was set up an artistic and sensory journey through the squares of the city, with five installations by young designers, dedicated to the touch, sight, taste, hearing and smell.

第四届 Ecole del Rusco 主题活动：这次关于艺术和回收利用的展览在博洛尼亚举行，一次艺术和感知之旅在城市的几个广场展开，年轻设计家以五个艺术装置去表现触觉、视觉、味觉、听觉和嗅觉。

The temporary art installation "Circolare" is a project of landscape using products made with recycled materials compared with the same materials in their original form. The dialectical opposition between the waste raw material in its original form and the object made from it has the dual function of showing concrete results from a recovery action and stimulating the tactile curiosity of the public who can verify personally the effect of such recovery. The containers that collect the protagonists of this dialogue are big spheres that symbolically through their shape indicate the circular process of the recycling act.

临时的艺术装置 Circolare 是一个景观工程，使用了那些不再是原有形式的回收材料做成，在原始形式的废弃原材料与用其制造的东西之间形成了辩证对比。这有双重功能，既展示了回收行动的具体成果，又刺激了公众的触觉好奇心，他们能亲自证明这种回收的效果。这种交流的主角放在巨大的球体里，它们的外形象征着回收行为的循环过程。

The building of the new municipality stands in the middle of a neighbourhood like a giant steel and glass come from somewhere.The building does not belong to the landscape as well as the large spheres which appear to be rained down from another world.

新的市政大楼矗立在一个区域的中心，像从某个地方冒出的一个巨大的钢筋和玻璃体。大楼以及巨型球体都不属于景观，球体看起来像从另一个世界如雨点落下来的一样。

Behind them, on the lawn, the trail underscores the imaginary landing of these rolling meteorites.Three are transparent, covered with thin pencil-written, three black, made of a thick rubber like that of car tires.Latter contain the raw materials they are made of objects contained in the transparent spheres.

在球体后面的草坪上，尾迹让它们像陨石滚落下的感觉更逼真。三个球体是透明的，上面有纤细的铅笔字迹，三个是黑色的，由像汽车轮胎那样的厚厚的橡胶制成。后面提到的三个装着原材料，透明的三个里装的东西是由这些原材料做成的。

Columbusplein 广场
Public Square Columbusplein

LOCATION:
Amsterdam, the Netherlands
AREA:
9,400 m^2

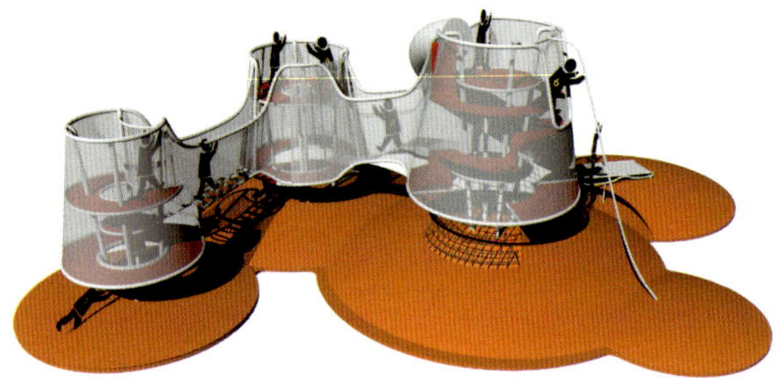

项目地点：
荷兰阿姆斯特丹
面积：
9400 平方米

Columbusplein is a public square in Amsterdam-West originating from the 1920s and typical for the 1920-40's urban layout of the city. Recently this area undergoes extensive urban renewal. Social housing is renovated and partly sold to stimulate diversity among the inhabitants – many belong to low-income groups and have immigrant backgrounds. In the past years social tensions were manifested by threatening behavior of groups of youngsters in this area, disturbing "normal" use of public space by divers groups.

Columbusplein 是位于阿姆斯特丹西区的一个公共广场，源自 20 世纪 20 年代，是 20 世纪 20 至 40 年代城市规划的典型代表。最近这个地方进行了大面积的城市改造。社会保障房进行了修整，部分卖给了居民以激发城市的多姿多彩——许多人属于低收入人群，有移民的背景。在过去几年里，这个地方成群的年轻人威胁性的行为表明了社会的紧张，妨碍了各类人群对公共空间的正常使用。

Situation

The elongated square had been divided in two parts mid 1980's, the two parts developed differently in use and atmosphere. The south was green, a secluded park with a hidden square inside that gave rise to undesired forms of use and drug and alcohol related hindrance, and made it an avoided area. In a disorderly layout in the north, worn out playground equipment was combined with an undefined asphalt-area for sports. This side of the square was frequently used by an adjacent primary school, and the children had to cross the road.

情况

这个狭长的广场在 20 世纪 80 年代中期已被分成两部分。这两部分在功能和氛围上有所不同。南面是一个绿色的僻静公园，里面有一个隐蔽的广场，这里常有吸毒、酗酒等妨碍社会的行为，成为一个人人避之不及的场所。在北侧过去布局混乱，破旧的操场设施连同一个不成形的沥青场地用于运动。广场这一侧过去被一个毗邻的小学频繁使用，孩子不得不跨过街道去玩。

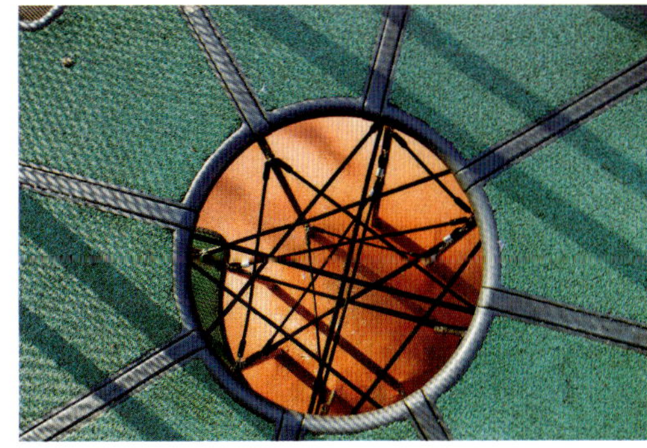

亨克城市文化公共中心
Genk C-m!ne - Hosper

LOCATION:
Genk, Belgium
AREA:
0.5 hectares
DESIGNERS:
Hanneke Kijne, Petrouschka Thumann, Marike Oudijk, Remco Rolvink, Ronald Bron, Hilke Floris, Han Konings
PARTNERS:
ARA Atelier Ruimtelijk Advies, Carmela Bogman industrial design
AWARDS:
first prize design competition cultural square

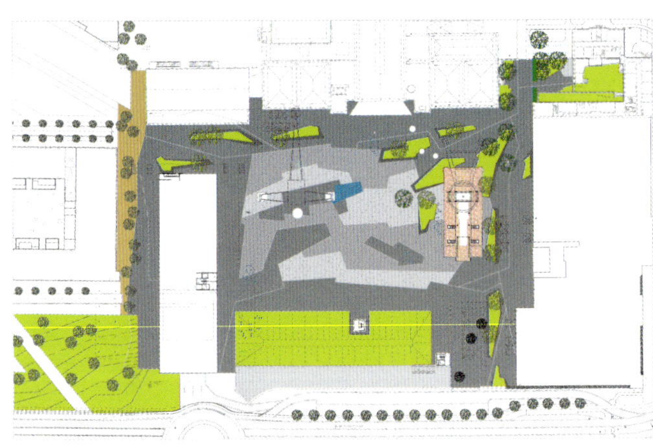

项目地点：
比利时 Genk
面积：
0.5 公顷
设计师：
Hanneke Kijne, Petrouschka Thumann, Marike Oudijk, Remco Rolvink, Ronald Bron, Hilke Floris, Han Konings
合作伙伴：
ARA Atelier Ruimtelijk Advies, Carmela Bogman industrial design
奖项：
文化广场设计一等奖

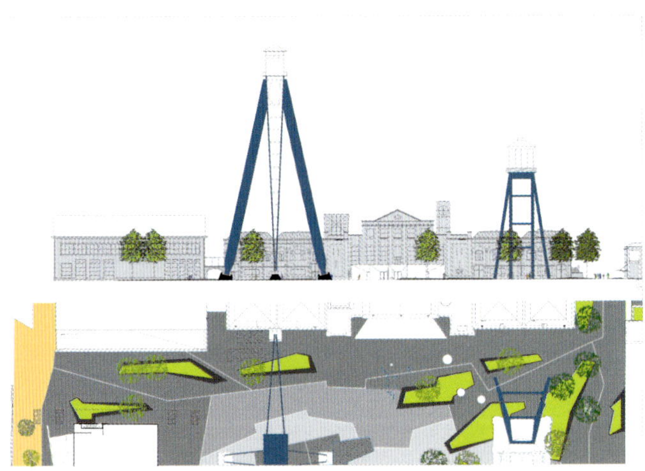

The C-M!ne square, situated on a former coalmining site, is the central open space of the new cultural center of Genk. It will become an urban square with a cultural, creative, design and recreational function. Most of the buildings around the square are former mining buildings, renovated and transformed into buildings with a cultural program; a large theatre, a cinema, restaurants and the (newly built) design academy of Genk. The design of the square interacts with the surrounding buildings and will facilitate and create space for all sorts of spectacle. The square makes a spectacular open space; the events and activities planned on the square enhance the square as the cultural heart of Genk.

坐落在早先是采煤场的一个场地的 C-M!ne 广场是 Genk 新文化中心的开放地带。它会成为一个兼具文化、创意、设计和娱乐功能的城市广场。广场周围的建筑物大多是早先的采矿建筑，后来为了一个文化项目而装修、改造成了大剧院、电影院、餐馆和（新建的）Genk 设计学院。广场的设计与周围的建筑映衬，将会为各种活动提供便利条件和创造空间。广场是个壮观的开放空间，广场上的活动使其成为 Genk 的文化中心。

The need to create opportunities for a wide range of activities on the square results in a "one level surface". An obstacle-free surface ensures that the square can be used for a wide variety of purposes. Of course, at times of activities and a large numbers of visitors the square will be lively and marvellous. However, it will remain a very special square even when there are fewer visitors, no activities and the surrounding buildings are outside normal opening hours.

“一个平面”满足了广场上为广泛的活动创造机会的需要。一个无障碍的表面确保广场可用于各种各样的用途。当有活动的时候，广场会聚集大量的人，当然也就会充满活力。而即使在没有多少访客，没有活动，周围建筑也都关门的时候，它仍然很特别。

The square is paved with black limestone slabs of different sizes and laid in an informal pattern. The black limestone refers to the "black gold" from the mines. The paving includes lighting in the surface as well as the possibility for a water surface, the creation of mist just above the surface and removable seating. A great deal of attention is given to the night-time appearance of the square, with lighting illuminating the surrounding facades and the mining shaft towers. The two shaft towers are both given a function on the square. The Belgian office Nu architectuuratelier designed an attractive route that follows the former mining corridors under the oldest shaft tower and the ruin of the former mining reception building and ends with a fantastic view on top of the youngest and tallest shaft tower.

广场是用不同大小的黑色石灰岩砖以不规则的图案铺成的。黑色的石灰石，是指矿山的“黑金”。铺路，包括在表面的照明，水面的考虑，表面上的雾气和可移动的座椅。广场夜间的外观得到了极大的关注，灯光照亮了周围的外墙及采矿塔。两座塔都在广场上发挥了功能。比利时的 Nu architectuuratelier 设计了一条吸引人的路线：沿着最古老的塔下面的昔日的采矿走廊和接待大楼的废墟，最终可以在最新且最高的塔的顶部一览美景。

Carmela Bogman, working in cooperation with HOSPER, has designed special seating for the C-m!ne leisure zone in Genk. The furniture is concentrated in a number of "clouds" on the square, where the chairs and stools are arranged in an informal pattern. Different arrangements are thus possible for people who want to sit close together and for people who prefer to sit further apart, facing one another or with their backs to one another. The chairs and stools, which are made of a folded stainless steel plate, glitter like diamonds against the black surface of the square. The internal and rear surfaces of the furniture are powder-coated in fire-engine red. The red surfaces of some of the chairs are lit from below, thus creating a warm glow around them at night.

与 HOSPER 一起合作的 Carmela Bogman 已经为 Genk 的休闲区 C-m!ne 设计了特殊座位。座位主要集中在广场上的“云”上，在那里椅子和凳子随意摆放。各异的摆放可以使想坐在一起的人们坐在一起，想坐得远的背对而坐。在广场黑色表面的映衬下，折叠式的不锈钢板做成的凳子和椅子像宝石一样闪闪发光。坐具内部和后部的表面涂上了消防车般的红色涂料。因此从下面照亮的一些椅子的红色表面在夜间营造了一种暖光效果。

C-m!ne is designed to be Genk's cultural center, with plenty of room and practically unlimited scope for all kinds of events and activities. One of the main requirements on the seating is therefore that it should be easily removable. To this end, each chair or stool is mounted with the aid of four bolts on a base, which is designed so as to blend into the rest of the road surface without leaving any differences in level over which people could trip when the seating has been removed.

C-m!ne 有着充足的空间，活动的规模几乎不受限制，将会成为 Genk 的文化中心。座位的主要要求之一就是要可以轻松地拆卸掉。为此，每个椅子或凳子都在原有的基础上安装了四个螺栓，这是为了当人们走了座位被拆除后，与其他路面不留差异地融合在一起。

Getsewoud Zuid 校园广场

Schoolyard and Playground Getsewoud Zuid

LOCATION:
Getsewoud Zuid the Netherlands
AREA:
7,000 m^2

项目地点：
荷兰 Getsewoud Zuid
面积：
7000 平方米

The schoolyard of the "brede school" Getsewoud did not function as intended. An immense building, housing three primary schools was realized only five years prior to our assignment. Situated in a recently build suburb of Haarlemmermeer, most public space surrounding it was already dedicated to infrastructure for commuters.

Getsewoud 社区学校的校园没有按设想的那样起作用。一座容纳三个小学的巨大建筑就在我们这一项目的五年前建成。它位于 Haarlemmermeer 市新建的一处郊区，周围大多数的公共空间已用于上班通勤所需的基础设施。

The main problems in the former situation concerned the bringing and picking-up of about 900 children in a very short time span resulting in a recurrent traffic jam; a lack of suitable scale and intimacy for young children; the costs for maintenance; and playable sculptures that were not used.

以前的情况涉及到的主要问题有：短时间内接送约 900 名孩子不断造成交通阻塞；小孩子缺乏适宜和亲密接触的空间；维护费用；供玩耍的雕塑没被使用。

To adjust the scale of the schoolyards on both sides of the building, we decided to divide the terrain into three parts by introducing a line of trees with benches. The benches can be used from both sides and function as an informal boundary.

为了调节楼两边校园的空间安排，我们设计了一排树，带有长凳椅，把这块地方分成三部分。长椅两边都可以使用，并且起到了类似边界的作用。

The now smaller yards for the younger children are made convex, up to a meter high, to create an inviting micro world. Borderless sandpits are sunken into the ground and together with the bridge that spans over them, they are a part of a "Manhattan-like" street grid in brick, which is also used for traffic education.

较小的场地是给年龄更小的孩子用的，地面凸起约一米高，为了营造一个吸引人的小小世界。地面挖有无边界的沙坑，上面跨有小桥，和地面砖块铺成的网格一起构成曼哈顿街道般的格局，可以用作交通教育。

The schoolyards for the older children are almost kept free of obstacles to allow for as much free space as possible, for all sorts of games. At one side, a vertical piece of play equipment incorporates a soccer goal, a slide and enough space for 30 children to play at the same time.

年龄大的孩子所用的校园几乎没有障碍物，为的是尽可能留出自由空间，可以进行各种各样的游戏活动。在一边上有一个垂直的游乐设施，包含一个足球门、一个滑梯和足够 30 个孩子同时玩耍的空间。

We convinced the city that here was a unique possibility to combine the schoolyard with the public playground situated next to it, offering a possibility to make a bigger area, dedicated to the children in this dense built "boring" suburb. It now also contains a multipurpose ball court with banked sides and ledges for skaters. The specially designed play equipment is placed on steep convex and concave safety surfacing to magnify the play experience dramatically.

我们让这个城市相信这里是唯一可能把校园和相邻的公共操场结合在一起的地方，能创造出一个更大的地方，专门为这个稠密而枯燥的郊区里的孩子所用。现在它还包含了一个多功能的球场，带有为滑板运动者使用的斜边和架子。这个特殊设计的玩耍设施被放置在忽高忽低的安全性地面上，明显增强了的玩耍的体验。

查普尔特佩克公园喷泉广场

Fountain Promenade at Chapultepec Park

LOCATION:
Mexico City
AREA :
3,480 m^2
DESIGN UNITS :
Grupo de Diseño Urbano
AWARD:
Honor Award in General Design. American Society of Landscape Architects

项目地点：
墨西哥城
面积：
3,480 平方米
设计单位：
Grupo de Diseño Urbano
奖项：
2008 年获美国景观建筑师学会总体设计荣誉奖

The major interventions in the rehabilitation of the second phase of Chapultepec Park are: Tamayo Park (12 hm^2) and Gandhi Park (8.2 hm^2.) As outlined in the original Master Plan, the intention was to reinhabit and attract families and users to this otherwise under use area. The strategy was through specific and discrete interventions such as: a fountain promenade (250 meters long by 20 meters wide) connecting the National Museum of Anthropology with the Tamayo Contemporary Art Museum, creating a new pedestrian axis-"paseo" through the park.

查普尔特佩克公园复原项目二期的主要工程包括：塔玛约公园（12 平方公顷）和甘地公园（8.2 平方公顷）。根据最初的总体规划，其意图是吸引居民和用户回迁到本地区，否则这里就利用不足。所采取的策略是建立一些具体、独立的设施，如修建一个喷泉步行区（250 米长，20 米宽），将国家人类学博物馆和塔玛约当代艺术博物馆连接起来，创建一条横穿公园的新的行人散步道。

The fountain cascades and water travel through existing trees were incorporated as part of the simple design geometry. Additional comfortable seating designed by the authors was incorporated to the promenade,creating a wonderful urban oasis of rest and Enjoyment.

喷泉瀑布和穿越现有树林的水渠组成了场地简单几何形状的一部分。设计师额外设计的舒适座椅摆设于散步广场之中，创造了一片奇妙的城市绿洲，供人们休息和娱乐。

水镜广场
Le Miroir d'eau

LOCATION:
Bordeaux, France.
AREA OF THE PLAZA:
3,500 square meters
FOUNTAINS AND WATER WORKS:
JML Consultants Water Feature Design (Jean Max Llorca, Stéphane Llorca)
ARCHITECTS:
Michel Corajoud with Atelier R (Landscape Architect), Pierre Gangnet (Architect)
PICTURES:
Stéphane Llorca – Haut relief
AWARD :
This project won the First Prize for new major urban space in France in 2008

项目地点：
法国波尔多
广场面积：
3500 平方米
喷泉和水利工程：
JML Consultants Water Feature Design (Jean Max Llorca, Stéphane Llorca)
建筑师：
Michel Corajoud with Atelier R (Landscape Architect), Pierre Gangnet (Architect),
图片：
Stéphane Llorca–Haut relief
奖项：
该项目荣获 2008 年度法国新主要城市空间一等奖

This project won the First Prize for new major urban space in France in 2008.The plaza is the heart of the new waterfront designed by Michel Corajoud.

这个项目在 2008 年赢得法国新主要城市空间一等奖。广场是 Michel Corajoud 设计的新滨水区的中心

In front of the stock exchange building, a unique example of classical French architecture, beside the Garonne river, the Plaza has an area of 3,500 square meters. The water skin is covering 80% of this area.

广场位于法国古典建筑独一无二的典范股票交易大楼的前方，加伦河边上，占地 3500 平方米，水面占了广场的 80%。

It offers a unique sensory experience to interact with water in an urban context. It is surprising and unexpected to see the space transforming into a large shallow pool.The Plaza is periodically flooded and drained in minutes. A series of misters are then activated to create an enchanting atmosphere.

在城市的背景里与水进行互动提供了一种独特的感官体验。看着这个空间转变成一个巨大的浅水池会觉得吃惊和出乎意料。广场定期在几分钟内注水和排水，接着，一系列喷雾器被激活以营造醉人的氛围。

This project blurs the lines between nature and architecture. The buildings are reflected in the perfect liquid mirror, while the water merges with the perspectives of the river in the background.

该项目模糊了自然与建筑之间的界限。建筑物被倒映在这个完美的水镜里，同时水与远处的河景融为一体。

When the plaza is flooded it becomes an endless playground for kids. The public understands very quickly the cycle of the water: flood, drain, mist, flood, etc. The 2cm of water are enough to wet the feet, or get totally soaked in. This open space offers an invitation to kick off your shoes and cool down in summer by "walking on water".

当广场注水时，它就变成一个无边的儿童操场。公众马上就知道水的循环情况：注水，排水，水雾，注水，等等。两厘米深的水足够弄湿双脚，或是完全浸透。这个露天空间可以让人漫步水上，吸引你踢脱鞋子，在夏日里感受凉爽。

The water is stored in a large tank located in the plant room. The system is computerized and automatically runs the programmed sequences. The water quality is permanently monitored ensuring good hygiene conditions.This project has opened a new door in terms of water feature design. The concept is a perfect balance of simplicity and complexity and offer myriads of variations.

水被储存在机房巨大的水槽里。该系统由计算机控制，自动运行编好的次序。水质一直被监测，以确保良好的卫生条件。该项目在水景设计方面进行了创新，设计理念是简单与复杂的完美平衡，制造出无数变化。

河边壮观的公众广场

Maaskade Cuijk

LOCATION:
Cuijk, Netherlands
AREA:
11,000 m²
LANDSCAPE ARCHITECTURE:
Buro Lubbers (in cooperation with Ballast Nedam)
PHOTOS AND TEXT:
Buro Lubbers
DESIGN COMPANY:
Buro Lubbers landscape architecture and urban design

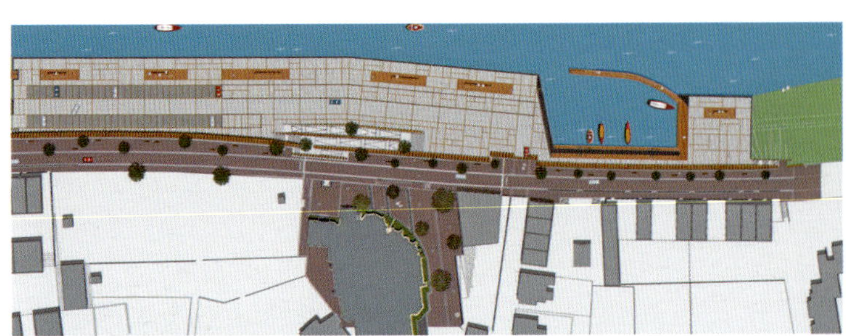

项目地点：
荷兰 Cuijk
面积：
11 000 平方米
景观设计师：
Buro Lubbers (in cooperation with Ballast Nedam)
图片和文字：
Buro Lubbers
设计公司：
Buro Lubbers landscape architecture and urban design

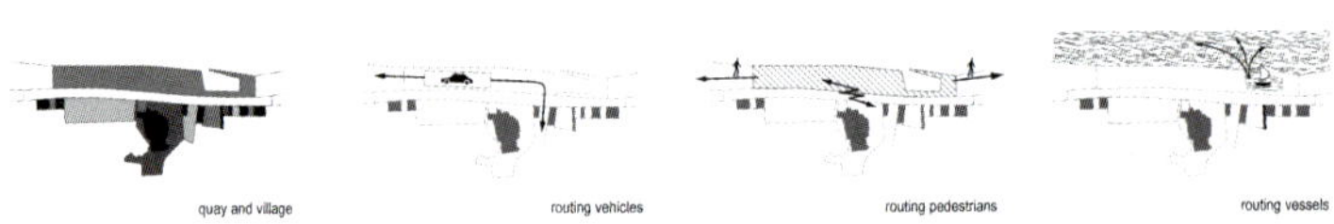

Originated in Roman times, the village of Cuijk can look back at a long history on the Meuse river. Until the fifties of the twentieth century, Cuijk was located directly at the water front, merely protected against high water by a wall. In the fifties a dike was constructed which consequently forced buildings at the Meuse to give way. A main road on top of the dike further cut of the river from the center of the village. A huge barrier between the village center and the river with its scenery was the result. Buro Lubbers designed a stunning intervention that reconnects both locations. The new embankment has become a pleasant public space where people can meet and events can take place.

墨兹河边的 Cuijk 小镇起源于罗马时代，历史悠久。Cuijk 直到 20 世纪 50 年代都直接位于河边，仅靠一道墙保护，阻挡河水。在 50 年代建起了一个堤坝，因此河边的建筑物被迫腾出地方。堤坝上的一条干路把河和村子中心进一步拉开。结果在村子中心和河之间有了一个巨大的带有风景的屏障区域。Buro Lubbers 对该地带做了一个令人震撼的设计，把两边重新连接在一起。新的堤岸已经变成一个怡人的公共空间，人们在那里可以聚会和举办活动。

Sturdy design
The design concept of the river bank is inspired on the scale and atmosphere of the river. The sturdy wall of steel has become the link between village and river, not by separating the two, but instead by giving both land and water more space. This effect is created by banking the wall 10%, by seemingly bulking it out of the river bank. Herewith one experiences a fluent and spatial transition between quay and water, between village and landscape.

坚固的设计
该河岸的设计理念是被河的规模和魅力激发出来的。那道坚固的钢铁墙现在变成村子和河流之间的连接，这一连接不是把两者分离开，而是给两边更多的空间。这个效果是通过让墙有 10% 的倾斜度来实现的，看起来好像墙是从河岸鼓出来的。于是人们体验到码头与水面、小镇与景观之间流畅的空间过渡。

The quay itself is leveled and designed as a plateau of stelcon plates with border lines of corten steel. Whereas the lines provide a subtle layout, the concrete panels show the contours of Roman archaeological finds, such as canals, a bridge pillar as well as explanatory texts.
The connection between quay and village center is first of all created by an underpass of the dike. Besides, near the church, a combination of ramps and stairs is realised, that connects both locations. Via the ramps one can slowly descend to the quay, the intermediate steps offer a quick link between the riverside and the top of the bank. Both ramps and stairs function as a viewing gallery during events, such as the Four Days Marches of Nijmegen.
Along the waterfront wooden platforms with furniture are placed. Here visitors can quietly overlook the river. A surprising feature is that the existing port is integrated in the design of the quay by applying the same wooden materialization. The wide wooden arm embraces the river and offers a nice panoramic view.

码头表面平坦，被设计成一个铺着 stelcon 板的高平台，以耐候钢为边线。这些边线界定了一个精美的布局，但混凝土平面上显现了罗马建筑遗迹的轮廓，比如运河、桥墩以及说明文字。
码头与小镇之间的连接首先是由堤坝的地下通道实现的。而且，在教堂附近，修建的斜坡与阶梯共同连接了两边区域。通过斜坡，人可以慢慢下到码头，中间的台阶把河边与岸上快捷相连。斜坡和阶梯在尼美根四天游街等活动时起到了观景长廊的作用。
沿着水边放置带有设施的木头平台。在此参观者可以安静地眺望河面。一个让人惊叹的特色是现有港口使用了同样的木质造型，与码头设计相容。宽阔的木质臂弯拥抱着河水，让人欣赏到美丽的全景。

Technical challenge

Because of the location at the riverside, the design was determined by several technical conditions. The essence of the intervention was of course its function as dyke, preventing the Meuse from covering Cuijk. In today's tempestuous climate, the struggle against water is increasingly important. Whereas the Meuse has an average elevation of +7.65 NAP, in times of high water it can state +13.50 NAP. Therefore the dam wall must have a minimal height of +15.00 NAP. The possibility of high water also had implications for the design of the elements in front of the wall, since these elements form obstacles in the river's stream in times of high water. That is why the ramps and the stairs are constructed in such a way, that the pressure and aspiration of the water does not damage their construction. The total construction is thus prepared with 200 mm reinforced concrete, a massive layer that is so strong that it can not float and can exist independently of its foundation. The other elements, like the wooden platforms, are detachable so they can be removed at high tides.

技术挑战

因为处在河边的位置，几个技术条件决定了设计。项目的核心当然是它的堤坝功能，防止墨兹河淹没 Cuijk 小镇。在当前多暴风雨的天气里，与水的抗争越来越重要。尽管墨兹河的平均水位是高出阿姆斯特丹水平面 7.65 米，但在高水位的时候，可达到 13.5 米。因此堤坝墙最低高度要高出阿姆斯特丹水平面 15 米以上。可能出现的高水位也影响了堤坝墙前面设施的设计，因为在高水位的时候，这些东西阻挡了河水的流淌。这就是那样建造斜坡和阶梯的原因，使水的压力和抽离不破坏建筑。整座建筑备有一个 200 毫米厚的巨大钢筋混凝土层，十分坚固，不会漂移，可独立于地基存在。其他的东西，像木头平台，都是可拆除的，在水位高时可被移走。

New landmark

Since 2008, the stunning intervention of Buro Lubbers reconnects the Meuse and the center of Cuijk. The carefully executed design is characterised by daring and sophisticated measurements. The huge quayside does not compete with its environment, but forms a subtle harmony. The church that rises high above the quay, the cargo-vessels that pass by on the river, the green river forelands and the discharge docks, all complete each other. While the quay at the side of the town emanates security and the wooden seats at the waterfront offer a grand view, the view at Cuijk from the opposite of the Meuse has changed dramatically. Not only Cuijk acquired a new meeting place and is protected against the unpredictable Meuse in the stormy weather of these days, above all the Dutch river landscape gained an explicit landmark.

新地标

自从 2008 年，Buro Lubbers 设计的这个让人叹为观止的项目就把墨兹河和 Cuijk 小镇重新连接在了一起。设计执行认真细致，特点是大胆的突破创新和精密的测量计算。巨大的码头区域与周边环境不是相互干扰，而是形成微妙的和谐。矗立在码头上方的教堂，河上驶过的货船，绿色的河岸和卸货船坞相互补充，彼此相容。小镇边上的码头提供了安全保障，坐在水边的木座上可以欣赏壮观的景色，同时从河对岸观赏到的小镇风光有了巨大的变化。不仅 Cuijk 小镇得到了一个新的公众活动场所，得到保护，免于当前暴风雨天气时墨兹河的不可预测带来的危险，更主要的是荷兰河岸景观又多了一个鲜明的地标。

Note: the quay side is the first phase of an urban and landscape project of Buro Lubbers in Cuijk. In next phases, among other things, the church and museum square and the cemetery will be redesigned and a plan will be formulated that defines the image quality of housing at the Maasboulevard.

注释：码头区是 **Buro Lubbers** 在 **Cuijk** 小镇所做的城市和景观项目的第一阶段。在接下来的阶段，除了其他工程，教堂与博物馆广场和墓地也将被重新设计，这里将制订一个计划，塑造起 **Maasboulevard** 住宅区的高品质形象。

Potgieterstraat 街

Potgieterstraat

LOCATION:
Amsterdam, The Netherlands
SITE AREA:
1500m^2
DESIGN:
Carve
TEAM:
Elger Blitz, Mark van der Eng, Renet Korthals Altes, Jasper van der Schaaf, Lucas Beukers, Stef van Campen

项目地点:
荷兰阿姆斯特丹
场地面积:
1500 平方米
设计:
Carve
团队:
Elger Blitz, Mark van der Eng, Renet Korthals Altes, Jasper van der Schaaf, Lucas Beukers, Stef van Campen

Small interventions often evoke bigger changes. Local involvement in a design for a street in the city of Amsterdam became the stage for public participation.

小的景观设计经常引起更大的改变。阿姆斯特丹市一个街道的设计有了当地人的加入成为了公众参与的舞台。

Previous state
Potgieterstraat is situated in inner Amsterdam, in a context of 19th century buildings dating back to the first big enlargement of Amsterdam. The block typology of that time appears to the disadvantage of today's public life, since the inner courtyards of these blocks are not open to public use and the streets were never designed for today's traffic. In general there is a lack of public squares and public green. Streets here are dominated by cars and recently introduced bike lanes are a traffic solution, unfortunately claiming the available space from adjacent side walks.

先前状况
Potgieterstraat 街位于阿姆斯特丹市内城，置身于可追溯到城市第一次大规模扩建的 19 世纪建筑背景中。那个时代的街区类型看起来给今天的公众生活带来弊端，因为这些街区内的庭院不对公众开放使用，街道根本不是为今天的交通设计的。总体来说，缺乏公共广场和公共绿色空间。车辆主导了这儿的街道，最近引入的自行车车道是交通问题的一个解决办法，但不幸的是，占了旁边人行道的可用空间。

The district as a whole was up to a refreshing new strategy for children and pedestrians to strengthen and vitalise the public realm. Local inhabitants were asked in a political enquiry to agree upon and formulate new guidelines and were also involved in the selection of an architect.

这个区域整体上看有条件为孩子和行人进行一次耳目一新的规划，以加强和活跃公众空间。政府要求当地居民协商制定新的方针，以及参与选择建筑师。

Aim intervention
Carve suggested to close down one of the streets entirely for car traffic in order to rededicate that street space to citizen utilization. The rededicated former street and parking area given to Carve to design, has a total site surface of 1,500 m^2. The site's functional program was changed from traffic and parking into an urban program of meeting and pausing places, a playground for children, an upgrading of the green quality and overall, a positive urban beacon for the district.

规划目标
Carve 建议把其中一条街彻底向车辆交通关闭，为了把这个街道空间重新给市民使用。这个以前的街道和停车场地被划出给 Carve 去设计，场地总面积是 1500 平方米。场地的功能从交通流线和泊车变成一个都市聚会和小憩的场所、一个孩童的操场、一次绿化质量的升级，总之，是该区域毋庸置疑的焦点。

Intervention

Carve's intervention was firstly to rethink the street into a play street, accessible only to bikes and pedestrians. All surface materials were removed, the existing trees however were kept and new ones added. Into that clearance, Carve designed a mogul landscape with play objects integrated, materialised in abstract black rubber. The play objects vary from interactive elements to water sprayers. The rubber can be used as a drawing surface, invites to jump, run, fall thanks to its soft feel while reducing noise levels.

规划

Carve 的规划首先是重新考虑把街道变成一个玩耍的地方，仅自行车和行人可进入。表面所有的东西都被拿开，而现有的树木保留下来，增加新的树木。清空后，**Carve** 设计了一个很有气势的景观，玩耍设施融入其中，结构外形是由抽象的黑橡胶制成的。玩儿的东西很多，从互动的设施到喷水器等等。橡胶表面可以用作绘画，它柔软的感觉吸引人在上面跳、跑、跌倒，同时降低了噪声。

However, the true benefit of this design is not obvious on a first glimpse. It is rather the reclaiming of local urban realm by its community. Parents but also citizens without children interact and relax here on wooden benches and around a little kiosk. The location becomes an anchor for neighbourhood interaction and interlocks as well its surrounding blocks as well as helping to get together people of different backgrounds and ages.

然而，这个设计真正的利处不是看一眼就明白的，关键是社区重新收回了当地这块城市空间。父母们还有没孩子的市民在小凉亭边坐在长木凳上交流和休息。这个地方成了街邻四坊交流的固定场所，也紧密连接了周围的街区，有助于把不同背景和年龄的人聚到一起。

van Beuningenplein 广场

van Beuningenplein

LOCATION:
Amsterdam, The Netherlands
SITE AREA:
9,500m^2
LANDSCAPE ARCHITECTURE:
Dike & Co.
SPORT AND PLAY:
Carve

项目地点：
荷兰阿姆斯特丹
场地面积：
9500 平方米
景观建筑：
Dike & Co.
运动和玩耍区：
Carve

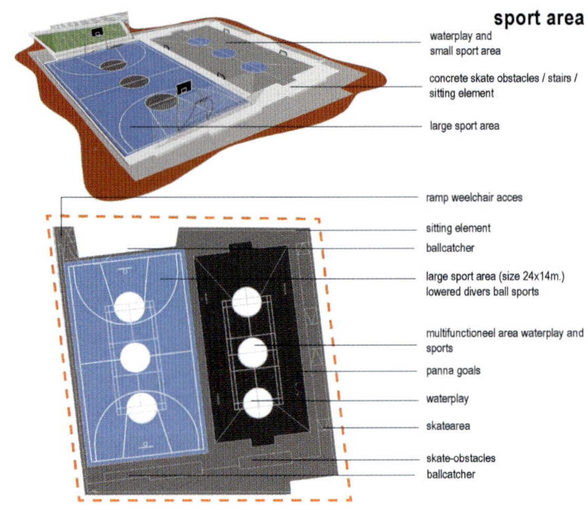

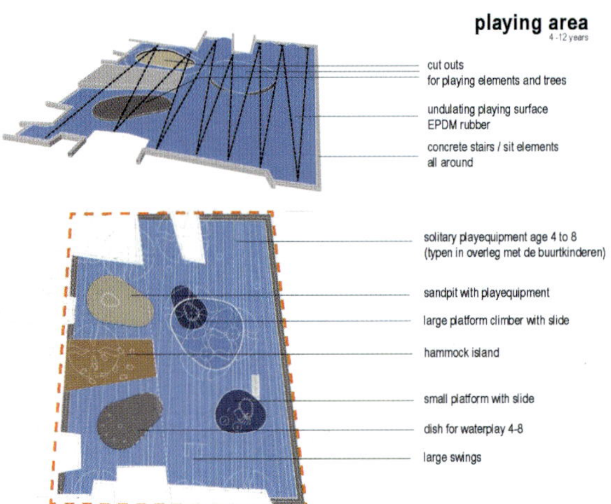

Parked cars dominate the public space both physically and functionally. Therefore the city of Amsterdam has decided to construct an underground parking garage at "van Beuningenplein" at the existing playground. On top of the new parking garage, the former play and sport area, had to return.

停泊的车辆在视觉和功能上主宰了这个公共空间。因此阿姆斯特丹决定在现有的 van Beuningenplein 广场下方建造一个地下停车库。在新的停车库上方，以前的玩耍和运动空间需要重现。

In the former situation the "van Beuningenplein" was hidden from view by cars, fencing and poorly maintained green. By eradicating the cars and other obstructions the facades of the surrounding houses are connected to the square and the square becomes once again part of the neighbourhood.

以前的情况是 van Beuningenplein 广场被车辆、栅栏和维护不善的绿树遮挡视线。把车辆和其他障碍物移除后，周围房屋的正脸与广场连接了起来，广场再次成为该生活区的一部分。

Along the facades hedges are placed in strategic locations leaving space for resident initiatives, like a bench or a facade garden. The boundary between private and public has become less rigid, a colourful and lively plinth is the result. Green borders of perrenials frame the central part of the square without isolating it from its surroundings.

沿着建筑物的正面，在关键位置放置了篱笆，为居住功能留出了空间，比如长凳或外层花园。一个色彩丰富和充满活力的基座让私人和公共空间的界限变得不那么死板了。多年生植物为广场中央筑起绿色边界，使中心与四周连接和谐。

The central part is designated for sports and play. On the sport field there is also space for a water feature and in wintertime ice-skating is possible. By placing special elements on the edge of the sunken sport field a skatable edge has been created. The playing area is a large wavy surface with different playing towers, play elements for all ages and a water playground for summertime.

中心部分专门用于运动和玩耍。在这个运动场地上，还给一个水池留出了空间，在冬天可以滑冰。在这个凹陷的运动场地边缘放置了特别的设施，可在上面做滑行。玩耍区域是一个巨大的波浪状表面，带有不同的玩耍区域、适合各年龄段人们玩的设施和一个用于夏日的水上操场。

The sports and play features are extended with a youthcenter combined with an exit and entrance of the car park, a small building for a playground manager, and a public tea-house. Mixing all these functions allows different groups of people and different ages to come together. The whole square plays an important role in the social cohesion of the neighbourhood, hence becoming much more than just a playground.

一个带有车库出入口的青年中心是对这些运动和娱乐设施的补充，还有一个操场管理者的小房子和一个公共茶馆。所有的这些功能合在一起让各年龄段的人们来到这里。整个广场在这个生活区的社会凝聚力上起到了重要作用，所以它不仅仅是一个操场。

安克拉治博物馆广场
Anchorage Museum Expansion

LOCATION:
Anchorage, Alaska, USA
AREA:
2 acres
LANDSCAPE ARCHITECT:
Charles Anderson Landscape Architecture
IN COLLABORATION:
David Chipperfield Architects, Alaska, USA
AWARD:
2008 WASLA Design Merit Award (Work in Progress)

项目地点：
美国阿拉斯加安克拉治
面积：
2 英亩
景观设计师：
Charles Anderson Landscape Architecture
合作：
David Chipperfield Architects, Alaska, USA
奖项：
2008 年 WASLA 设计奖

The new landscape for the Anchorage Museum Expansion (AME) lies in the heart of downtown Anchorage, and will serve as a public common and park, exhibition space, and recreational site for the city. Its dual nature requires a design that is at once flexible and dynamic, while also creating a strong sense of place and identity.

安克拉治博物馆扩建（AME）的新景观位于安克拉治市中心的心脏地带，将成为一块公共用地，是集公园、展览和休闲为一体的场地。它的双重本质要求设计既灵活又活跃，同时也要塑造起地方及个性感。

It was important to create an identity for the Museum that is evident and unique on the street, yet in keeping with the city's regional context. Reflecting the natural history of Alaska's south central region, the concept for the Museum's civic space emerged from the expansive mudflats and the deciduous birch forests around Anchorage. The landscape, filled with birch trees, encompasses the site with a semi-transparent screen. The grid, composed of multi-stem trees on the west side, becomes more open toward the building, with single-stem trees and increased spacing. This urban forest affords the park a sense of openness and visibility while also creating a striking presence at the street. Its distilled structure provides a continuous framework that will inspire and accommodate art and host a diverse set of celebrations and activities throughout the year. The AME streetscape design considered view corridors, connections to downtown anchorage and nearby parks, a transit stop, vehicular drop offs, pedestrian access throughout the year, and safety concerns.

为这个街边鲜明而独特的博物馆塑造一种个性是非常重要的，但还要与城市的区域风光保持一致。博物馆这个公众空间的设计想法反映了阿拉斯加中南部地区的自然历史，源于安克拉治周围广阔的泥滩和落叶桦树林。这个景观满是桦树，把场地围绕起来，形成一个半透明的屏障。多干的树种植在桦树网格的西侧，然后靠近大楼栽的是单干树，间距增加，网格变得更开阔。这片城市树林让公园感觉开阔和醒目，在街边惹人注目，气势非凡。精美的布置打造出一块延绵的框架，激发并蕴含艺术，全年可以举行不同的仪式和活动。安克拉治博物馆扩建街景设计考虑了街道景观长廊、与市中心和附近公园的连接、公交站点、停车区、全年行人进入和安全问题。

The plants of this park are mostly native to the region. The exceptions are the lawn used in the common and forest rooms, the flowering crabapples of the axial allee, and the long lines of annual flowers that are a key part of the sense of summer and tradition that permeates downtown Anchorage. The birches make a rare and vital urban forest, complete with an associated plant community, that along with the other elements make a simple almost minimalist design expression that is rooted deeply in ecology and natural history. The landscape is a truly fitting compliment to the regal and atmospheric museum building.

该公园的植物多数是当地的。例外的是树林空间里的草坪，中轴小径两侧的开花红果和长排的一年生花卉，这些花卉给人以夏日的感觉，体现了安克拉治的传统，遍布在市中心。这些桦树打造出一个稀有而必要的城市森林，具有附属的植物群落，连同其他的东西，表达出简约的设计风格，立足于生态学和自然历史。这一景观是对华丽和气派的博物馆建筑的匹配补充。

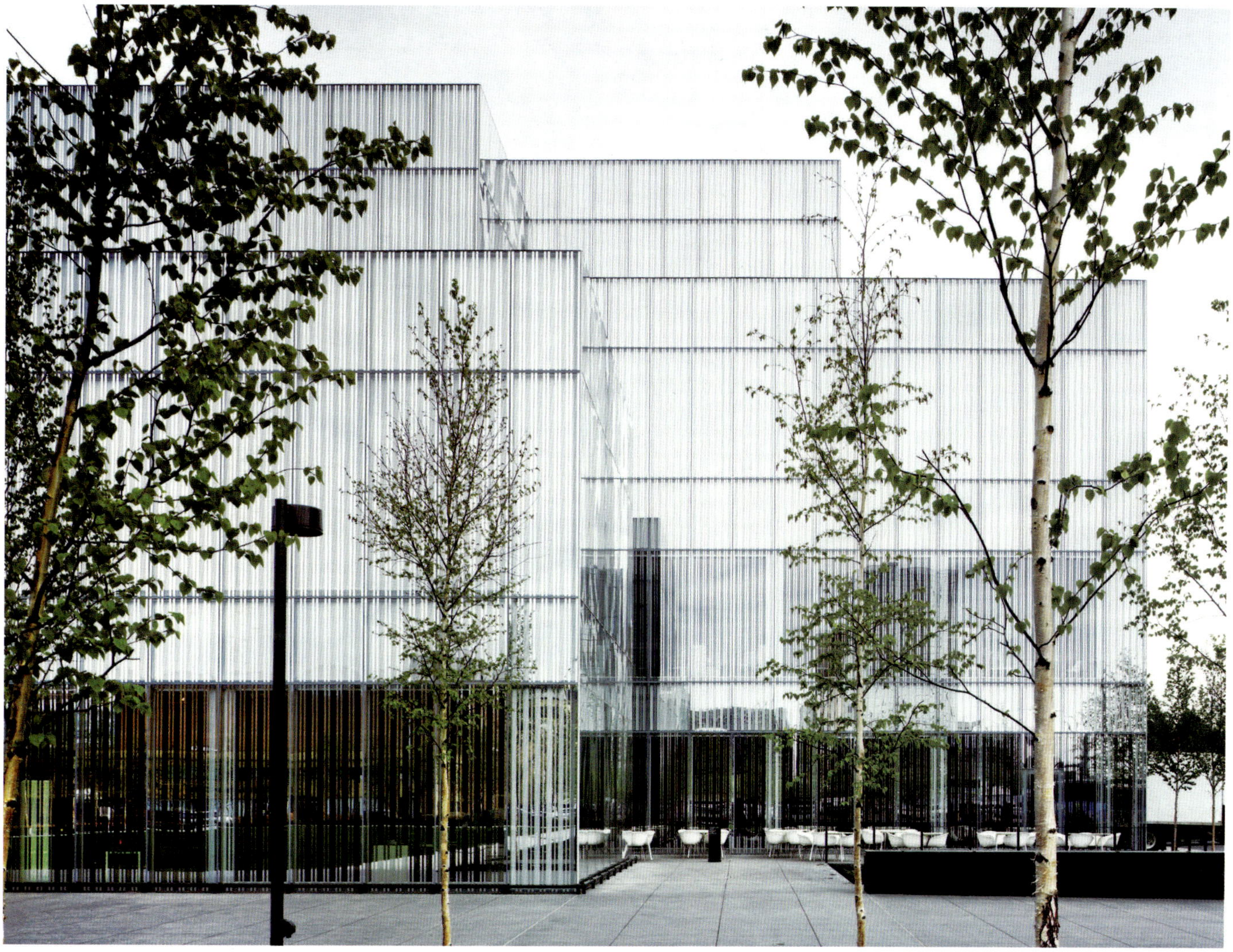

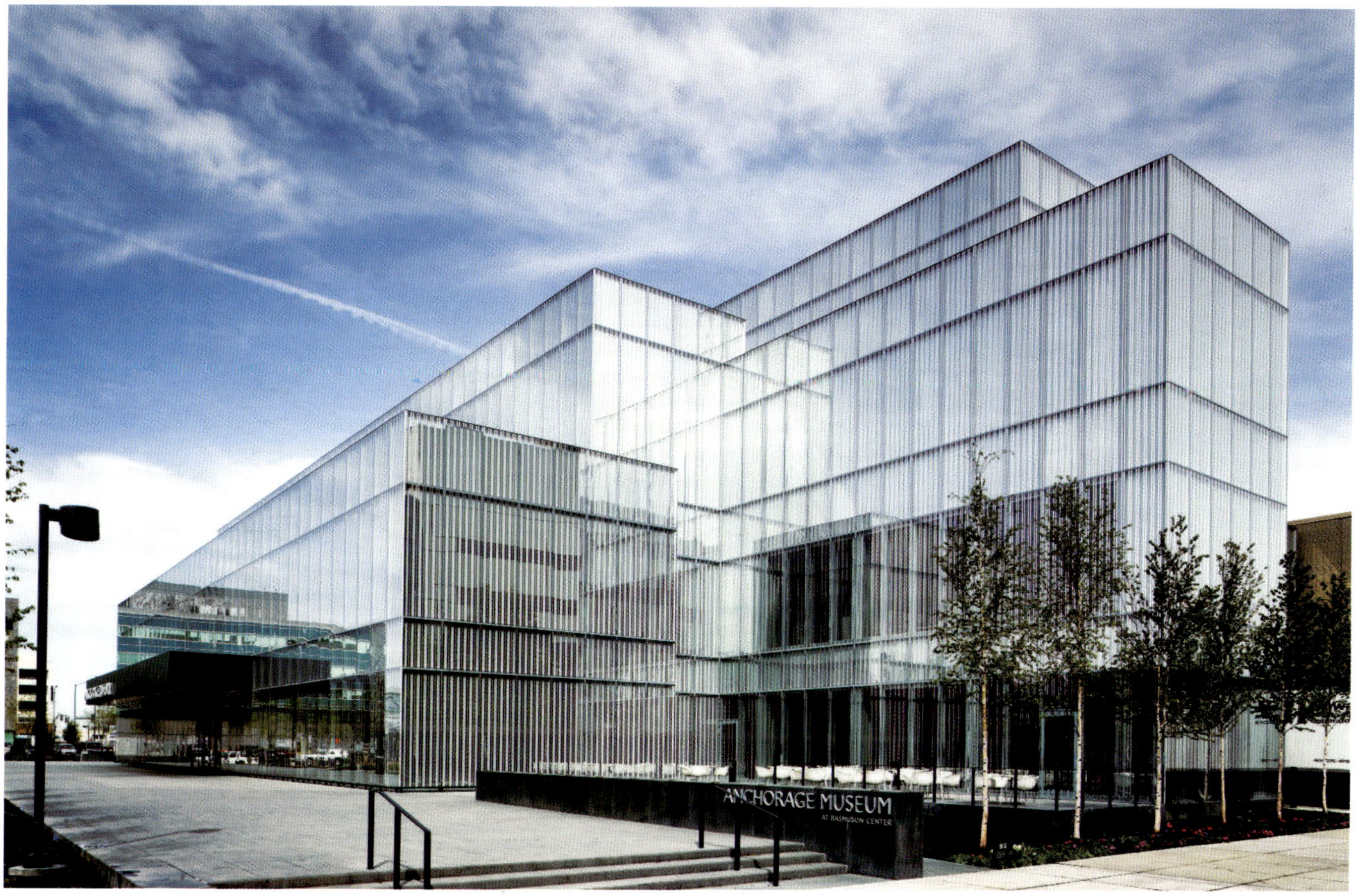
ANCHORAGE MUSEUM
AT RASMUSON CENTER

奥尔堡滨水区广场
Aalborg Waterfront

LOCATION:
Aalborg, Denmark
AREA:
170,000 m^2
ARCHITECT:
C. F. Møller Architects
LANDSCAPE:
C. F. Møller Architects, Vibeke Rønnow Landskabsarkitekter
ENGINEER:
COWI A/S
PRIZES:
1st prize in architectural competition, 2004

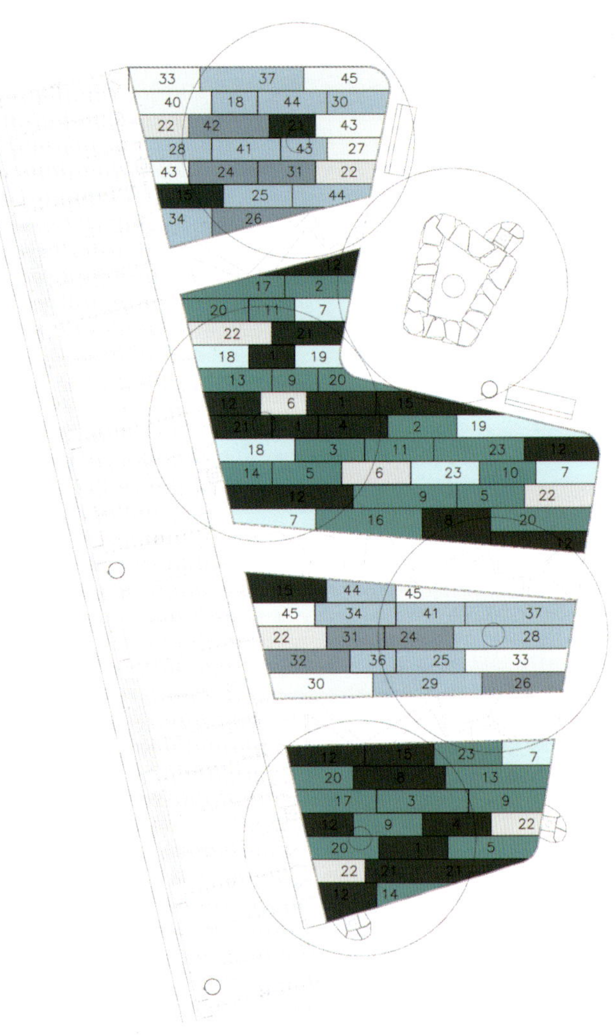

项目地点：
丹麦奥尔堡
面积：
170 000 平方米
建筑师：
C. F. Møller Architects
景观设计师：
C. F. Møller Architects, Vibeke Rønnow Landskabsarkitekter
工程师：
COWI A/S
奖项：
2004 年建筑设计竞赛一等奖

The master plan for Aalborg Waterfront links the city's medieval center with the adjacent fjord, which has previously been difficult for citizens to access due to the industrial harbour and the associated heavy traffic. By tying in with the openings in the urban fabric, a new relationship between city and fjord is created, and what was formerly a back-side is turned into a new, highly attractive front.

奥尔堡滨水区的总体规划把中世纪城市中心和毗邻的峡湾相连，峡湾因为过去的工业港口和连带的繁重交通让市民很难到达。通过与城市布局的接口处有机相连，城市与峡湾建立起新的关系，过去这里是一个城市的背侧，现在变成崭新迷人的一个前脸。

The qualities of the approximately one-kilometre stretch of quayside are emphasised with a tree-lined and unusually detailed boulevard to accommodate cyclists and pedestrians. The medieval Aalborg Castle once again becomes the harbour's centrepiece through the establishment of an extensive green area to frame the historic embankments.

这个大约绵延一公里长的码头区域突出的特征是设计了两旁栽着树的林荫大道，以容纳那些骑自行车的人和步行者。大片绿色区域的建立勾画出这个历史堤岸的结构，使中世纪的奥尔堡城堡再次成为港口的焦点。

At the same time, Aalborg receives a harbour promenade with steps and recessed terraces, allowing people to get close to the water. Various kinds of urban gardens facilitate activities such as markets, ball games and sun-bathing. The aim is to create robust and attractive spaces to benefit many different users.

同时，奥尔堡获得了一个港口漫步区，有台阶和凹进的露台，让人们与水靠近。各式各样的都市小花园给诸如集市、球类比赛和日光浴等活动提供了便利，目的就是打造坚固迷人的多个空间，让不同的使用者受益。

The central activities field is designed to accommodate various games and sports, from beach-volley in summer to ice skating rink in winter, surrounded by dramatically angled netting and lighting masts. The adjacent gardens are a calm, slightly sunken green space with a dense planting of trees and flowers. A planned floating harbour bath will be located along the waterfront, next to "Elbjørn" – a former ice-breaker converted into a floating restaurant/workshop.

中央活动场地的设计是为了给各类比赛和运动提供空间，从夏日的沙滩排球到冬日的滑冰场，周围是成特别角度的渔网和照明桅杆。毗邻的花园是一个安静、略凹陷的绿色空间，里面有浓密的花草树木。一个计划中的漂浮港口水池将沿滨水区而建，挨着“Elbjørn”号——曾经的一艘破冰船，被改造成一个漂浮的饭店和工作室。

The materials chosen are as raw as the fjord itself, including asphalt, rubber, corten steel, concrete and wood, while at the same time containing subtle references to the sea through wavy pavement patterns – an architectural quote of the famous Copacabana beach promenades by Roberto Burle Marx.

所选用的材料都像峡湾自身一样未经加工过，包括沥青、橡胶、耐候钢、混凝土和木头，同时通过地面铺筑的波浪图案与海有着细腻的呼应——这是借鉴了 Roberto Burle Marx 所设计的著名的科帕卡瓦纳海滩步行区。

“线条漫步”广场
Line Dance Square

LANDSCAPE ARCHITECT:
NIP Paysage
WORK TEAM:
Michel Langevin, Emilie Bertrand-Villemure, Mathieu Casavant, Josee Labelle, Melanie Mignault, Patrick Morand- NIP Paysage
AREA:
700m^2
CLIENT:
Developpement d'Arcy-McGee
PHOTOGRAPHER:
NIP Paysage, Jim Verbert

景观建筑师：
NIP Paysage
工作团队：
Michel Langevin, Emilie Bertrand—Villemure, Mathieu Casavant, Josee Labelle, Melanie Mignault, Patrick Morand—NIP Paysage
面积：
700 平方米
客户：
Developpement d'Arcy—McGee
摄影师：
NIP Paysage, Jim Verbert

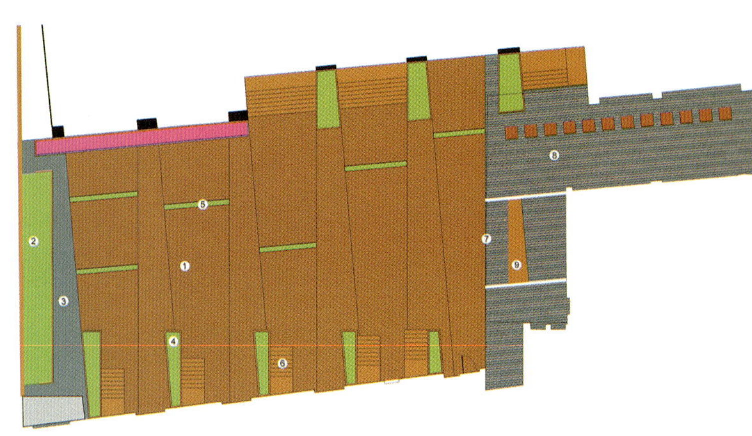

Unity courtyard is built on top of an existing parking garage, in between two luxury condominium buildings within Montreal's Quartier International neighbourhood.

团结庭院建立在现有的车库之上，位于蒙特利尔国际社区的两栋豪华公寓大厦之间。

Floor Pattern
© Jim Verbert

Based on the subtle dislocation of two architectural grids (between a restored industrial building and a new contemporary architectural piece), the design of the inner courtyard harmoniously links the rhythm of both facades in a dynamic and playful way. Wood is used throughout the project to contrast with the downtown "Hard Scape" and the resulting parquet motif recalls the slight shift between the two building footprints. Planters and steel shade structures create vertically framed spaces for both residential and commercial uses that share the space.

基于两个建筑网格（经过修复的工业建筑和一个新的当代建筑之间）的微妙的错位，内部庭院设计通过充满活力和趣味的方式和谐地连接两个外立面间的节奏。木材用于整个项目与市中心的“硬花篙”进行对比，由此产生的拼花图案记录了两座建筑物之间的微妙关系。花槽和遮阳的钢结构为住宅及商业共享空间创造了垂直框架的空间。

The heart of the courtyard comes alive with a large scale, perfectly flat, floor of western red cedar planks which fan out from one facade to another, offering diverse widths, uses and character. In the dancing motif of the wooden planks, the courtyard deck becomes the physical link and the meeting point between two architectural eras. Using the woody material creates a contrasting icon that distinguishes itself from the strong presence of brick and metal on adjacent facades and the cold concrete ambiance and feeling of nearby downtown Montreal. In a few strategic areas, river stones have been used as gabion walls to add texture to private planters and entrance features.

庭院的中心区域开始活跃起来，伴随着大规模的优美住宅的建立，西部红雪松木板以扇形从一个楼层到另一个楼层散开，提供多样化的宽度、使用方式和特征。在以舞蹈为主题的木板区，庭院木板变成了两座建筑之间的物理连接和交会点。使用木质材料创建了一个对比鲜明的外表，区别于相邻外墙的砖、金属、周围冷淡的混凝土氛围和附近蒙特利尔市的感观。在一些战略领域，河里的石头已被用来作为石笼挡土墙，从而增强私人花园和入口的特征。

Planters Detail
© Jim Verbert

Resting on the roof of the underground garage and surrounded by twelve floors of buildings, the courtyard offers some planting elements and structures to add greenery where conditions allow. These structures, resulting from the illusion of "Folding" of the floor surface, underline the verticality of the space and define private zones for residences that have direct access to the courtyard.

搁在地下车库屋顶上的休息空间和被 12 层高的建筑所包围，在条件允许的地方，庭院提供了一些可种植的元素和结构，以增加绿化植物。这些结构使人错认为是从地板表面产生的折叠结构，强调空间的垂直度，并为直接通往庭院的住宅区规划了私密空间。

Plant selection offers visual screening and more privacy near individual entrances. Vegetation within shared public spaces provide light shade and lush greenery with seasonal interest.

植物的选择提供了可视化的屏蔽和更多的私密性出入口。公共空间内的植被随着季节性交替提供浅色和郁郁葱葱的绿化带。

奥格斯堡 Wollmarthof 住宅广场
Augsburg Wollmarthof

AREA:
1,600 m^2
DESIGN COMPANY:
GESELLSCHAFT VON TOPOTEK 1
LANDSCHAFTSARCHITEKTEN MBH

面积：
1600 平方米
设计公司：
GESELLSCHAFT VON TOPOTEK 1
LANDSCHAFTSARCHITEKTEN MBH

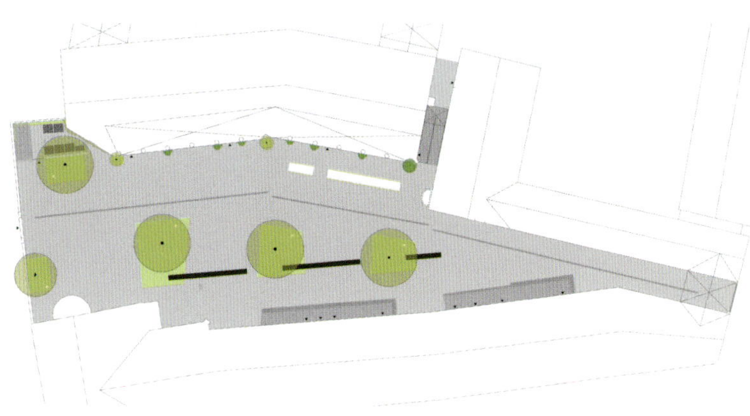

The Wollmarkthof is a heterogenic space, the different functions of the buildings exists next to each other without communicating with each other.

Wollmarkthof 是一个与众不同的空间：建筑的各个功能区都彼此相邻，但是彼此之间并没有联系。

The old cloister and the St Margareth Church are currently the only remaining pieces of a dispersed ensemble.

古老的修道院和圣玛格丽特教堂是目前历史遗留下来的唯一痕迹。

The main idea of our proposal is to make all the historic facades of the buildings visible and to solve the situation which emerged by the public and private use of certain areas, by the creation of an expanded public open square. An overall pattern emerged by the use of small sized paving stones which recovers the honor of the historic facades.

方案的中心思想是再现建筑的历史外观，并通过建造一个更大的公共广场解决某些区域在个人使用和公共使用中出现的问题。我们使用较小的铺路石打造出一个总体图案，再现了建筑历史外观的辉煌。

The surfaces of the used paving stones will be cut and assembled into a new mosaic pattern. Missing stones will be replaced by new ordered ones to create the desired image.

用过的铺路石会进行表面切割，然后拼成一个新的图案。丢失的铺路石要重新定购，以便能够打造出要求的图案。

The final image will compose all the different areas to one expanded surface.

最终的图案将把所有的区域都组合成一个展开的表面。

As an answer to the hedge two large, green marble benches will be placed along the front facade of the historic cloister. They provide a place to rest and also underline the entrance situation in front of the historic arcades.

古老修道院的正面设置了两个绿色的大理石长凳，与绿篱相互映衬，不仅提供了休息的地方，而且突出了古老拱廊前面的入口。

卢森堡 – 艾尔泽特河畔景观设计

Luxembourg by Aues Wird Gut Architektui ZT GmbH

LOCATION:
Luxembourg
AWARD:
Prix Luxembourgeois d'Architecture 2011 to announce

项目地点：
卢森堡
奖项：
2011 年卢森堡建筑大奖赛获奖

This house on a long, narrow and steep lot is centered around a big, sheltered courtyard with a mulberry tree in its middle. A double height space faces this outdoor patio and is revealed as a surprise to people entering the house. This arrangement blends the indoor and outdoor seamlessly, providing the family with a private garden all year round. The material palette for the walls and floors was intentionally limited to one type of natural stone, dressed in different ways.

住宅位于一个狭长、略陡的位置，周边是一个宽敞的庭院，院落中央有一株播撒阴凉的桑树。正对着户外的天井处设有一个双层高度的隔间，人们进入住宅每每大感惊讶。院落的布局让室内和室外浑然天成地融合在一起，给住户提供了一个全年的私家花园。墙壁和地板采用同一种材料——一种天然岩石，虽是同一材料，但最后呈现出的形态却完全不同。

SCHNITT C

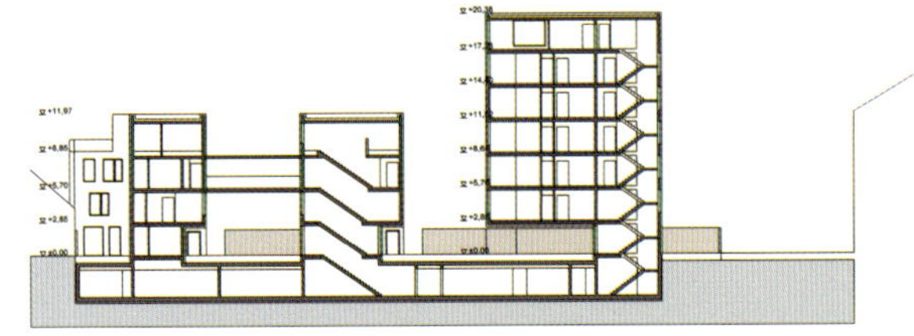

SCHNITT B

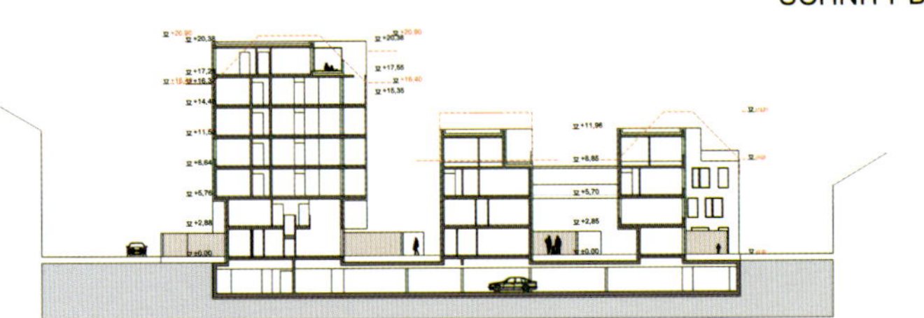

SCHNITT A

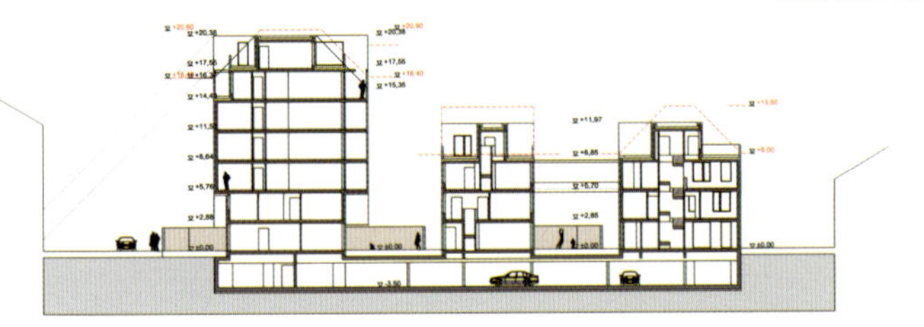

巴伐利亚国家博物馆广场
Bavarian National Museum

LOCATION:
Munich, Germany
AREA:
3,700 m^2

项目地点：
德国慕尼黑
面积：
3700 平方米

The design is a interpretation of the original plaza of Gabriel Seidel from 1898. The theme of the sunken forum was dealt with for the renewal of the square in front of the Bavarian National Museum. This theme, which already influenced the design of the city squares in the early 20th century, was taken up again.

本设计是对 1898 年 Gabriel Seidel 设计的原广场的诠释。广场（位于巴伐利亚国家博物馆前面）翻新采用了下沉式广场的主题，这一主题早在 20 世纪初期就已对城市广场的设计产生了一定的影响，现在新广场的设计又重新采用了这个主题。

The idea was to enlarge the plaza through the sloping levels, in order to change the perception of the viewer. The plaza becomes a new lifely and spirited space. The forum is framed by a homogeneous area made up of dark and light stripes that cover the square like a carpet. On two sides, framed with blocks out of granite, the plaza lowers regularly – beginning from the streetside (Prinzregentenstraße) – towards the main entrance.

设计思想是通过不同的倾斜平面扩大广场，为人们提供不同的感受，使广场成为一个充满活力和生气的新空间。广场采用了深色和浅色相间的均匀的条纹，像一张地毯一样，铺在广场之上，两侧是用花岗岩建成的体量。广场从街道一侧（Prinzregentenstraße 大街）向主入口有规律地下沉。

High quality materials such as white, grey and sand-coloured granite and cut Buxus sempervierens hedges shape the new design of the plaza.The dark green of the hedges are building a strong contrast to the white stripes of the paving stone and the stainless steel framed lawn-banquetts.

优质材料，如白色、灰色和砂土色的大理石以及锦熟黄杨绿篱塑造了新的广场设计。深绿色的绿篱与铺路石的白色条纹和不锈钢围成的宴会草坪形成强烈对比。

Before the spring of 1937, the dimensions of the forecourt were different: a big homogeneous square with a variety of elements like the sunken forum with the monument of the Prinzregenten, a raised terrace with groups of trees and the temple of Hubertus.

在 1937 年春季以前，博物馆的前厅还没有达到现在的规模：一个较大的广场，包括一些小品，如下沉广场，Prinzregenten 大街上的纪念碑，一个栽有一些树木的台地以及休伯特斯寺庙。

The concept of the lowered forum from 1936 is taken up and newly interpreted. On two sides, framed with blocks out of granite, the plaza lowers regularly towards the main entrance. Next to the function of a proportions defining element, the lowered forum offers the opportunity to rest at the steps and gives the main portal more presence through the additional slope towards the building.

从1936年开始，下沉广场的设计理念就已经得到应用，如今又得到了新的诠释。在两侧大理石体量的包围之下，广场有规律地向主入口下沉。除了衬托其他特色元素之外，广场还为人们提供了可以休息的台阶，同时，通往建筑的斜坡使主入口更加突出。

To give the Museum back an appropriate forecourt, the forum is framed by a homogeneous area made up of dark and light stripes that cover the square like a carpet.

为了重新为博物馆建造一个合适的前厅，广场采用深色和浅色相间的条纹，像一张地毯一样铺在广场之上。

Functional claims such as a cycle path as well as the emergency access road and the access of passenger cars are integrated on the pavement surface without any kind of possible distinction. The pavement echoes the stripes of the dark and light structure of the facade. The material used here is granite like for the surrounding pavements, although in a higher quality.

一些功能要求，如自行车道、备用通道以及客车通道都安排在人行道之上，没有加以区分。人行道与广场表面深浅相间的条纹相呼应。周围人行道上采用的是类似大理石的材料，但是质量要比大理石好。

The subject of the group of trees that form a roof of leaves is being interpreted in a new way. Outside the forum two groups of each four trees emphasize the main entrance and give the plaza a new proportion.

树木的叶子仿佛形成了一个屋顶，这一主题在这里以一种新的方式得到了诠释。两组木兰树（每组四棵）屹立在广场之外，不仅强调了主入口，而且在广场上形成了一个新的特色。

布拉福德城市公园广场
Bradford's City Park

LOCATION:
Yorkshire, UK
AREA:
2.4 hectares

项目地点：
英国　约克郡
面积：
2.4 公顷

Football pitches

Piccadilly Gardens (Manchester)

Peace Gardens (Sheffield)

St Marks Square (Venice)

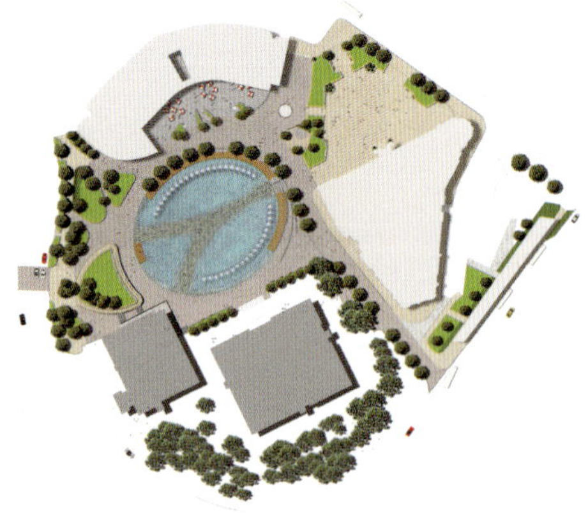

Bradford's City Park designed by Gillespies includes the UK's largest urban water feature and tallest city fountain.

Gillespies 设计的布拉福德城市公园包括英国最大的城市水景和最高的城市喷泉。

City Park in Bradford, designed by Gillespies landscape architects and urban designers on behalf of Bradford Council and supported by a multi-disciplinary design team is a landmark public space. It contains the largest city-center water feature anywhere in the UK, a 4,000 sqm mirror pool, and the UK's tallest urban fountain that reaches a spectacular 30 metres high.

布拉福德城市公园由 Gillespies 景观建筑师和城市设计师代表布拉福德市政设计，并得到了一个多专业设计团队的支持，是一处地标式公共空间。它拥有英国最大的市中心水景：一个 4000 平方米的倒影池和英国最高的城市喷泉，喷射高达 30 米，十分壮观。

City Park stems from a city center masterplan drawn up in 2003 for Bradford, which provided a vision of opening up the city center and creating a new public space. Bradford Council took the lead role in turning this vision into a viable plan, and Gillespies, Arup, Sturgeon North, Atoll and The Fountain Workshop developed this early concept into a detailed design which was submitted for planning permission and funding in 2007, before construction started in late 2009.

城市公园源于 2003 年起草的布拉福德市市中心总体规划，里面有开放市中心和建一个新的公共空间这样一个设想。布拉福德市政带头把这个设想变成了可行计划，然后 Gillespies, Arup, Sturgeon North, Atoll 和喷泉工作室将这个早期的想法变成了一个详细的设计，于 2007 年提交以获得规划许可和资金支持，在 2009 年年底开始建设。

The completed design has created a flexible dynamic centrepiece in the form of a vibrant 2.4 ha public space, including the mirror pool, fountains and public art. The park centers on Bradford's 19th Century City Hall and helps to connect the city's major visitor attractions with transport hubs and the rest of the city center. It enhances the overall image of Bradford and helps create a landscape for investment by setting Bradford apart from other cities.

完成的设计创造出一个具有灵活性和生命力的 2.4 公顷公共空间，是整个规划中最引人注目的作品，里面有倒影池、喷泉和公共艺术。公园是以布拉福德市 19 世纪的市政厅为中心的，帮助把位于交通枢纽地段的城市主要景点和市中心其他的部分连接在一起。它提升了布拉福德市的整体形象，使其与众不同，有利于创造投资环境。

图书在版编目(CIP)数据

风景园林. 2: 汉英对照 / 李壮主编; 吉典文化, 千朋万友编; 时真妹译. —大连: 大连理工大学出版社, 2014.8

ISBN 978-7-5611-9335-8

Ⅰ. ①风… Ⅱ. ①李… ②吉… ③千… ④时… Ⅲ. ①园林设计—世界—图集 Ⅳ. ①TU986.2

中国版本图书馆CIP数据核字（2014）第162202号

出版发行：大连理工大学出版社
（地址：大连市软件园路 80 号　邮编：116023）
印　　刷：深圳市新视线印务有限公司
幅面尺寸：220mm×300mm
印　　张：19
出版时间：2014 年 8 月第 1 版
印刷时间：2014 年 8 月第 1 次印刷
总 策 划：周群卫
责任编辑：张昕焱
封面设计：周艳丽 王志峰
责任校对：杨宇芳

书　　号：ISBN 978-7-5611-9335-8
定　　价：328.00 元

发　行：0411-84708842
传　真：0411-84701466
E-mail: 12282980@qq.com
URL: http://www.dutp.cn